SpringerBriefs in Computer Science

SpringerBriefs present concise summaries of cutting-edge research and practical applications across a wide spectrum of fields. Featuring compact volumes of 50 to 125 pages, the series covers a range of content from professional to academic.

Typical topics might include:

- A timely report of state-of-the art analytical techniques
- A bridge between new research results, as published in journal articles, and a contextual literature review
- A snapshot of a hot or emerging topic
- An in-depth case study or clinical example
- A presentation of core concepts that students must understand in order to make independent contributions

Briefs allow authors to present their ideas and readers to absorb them with minimal time investment. Briefs will be published as part of Springer's eBook collection, with millions of users worldwide. In addition, Briefs will be available for individual print and electronic purchase. Briefs are characterized by fast, global electronic dissemination, standard publishing contracts, easy-to-use manuscript preparation and formatting guidelines, and expedited production schedules. We aim for publication 8–12 weeks after acceptance. Both solicited and unsolicited manuscripts are considered for publication in this series.

This series is indexed in Scopus.

More information about this series at http://www.springer.com/series/10028

Tingting Yang · Xuemin (Sherman) Shen

Mission-Critical Application Driven Intelligent Maritime Networks

 Springer

Tingting Yang
School of Electrical Engineering
and Intelligentization
Dongguan University of Technology
Dongguan, Guangdong, China

Xuemin (Sherman) Shen ⓘ
Department of Electrical
and Computer Engineering
University of Waterloo
Waterloo, ON, Canada

ISSN 2191-5768   ISSN 2191-5776 (electronic)
SpringerBriefs in Computer Science
ISBN 978-981-15-4411-8   ISBN 978-981-15-4412-5 (eBook)
https://doi.org/10.1007/978-981-15-4412-5

This Springer imprint is published by the registered company Springer Nature Singapore Pte Ltd.
The registered company address is: 152 Beach Road, #21-01/04 Gateway East, Singapore 189721, Singapore

Preface

The development of maritime services has driven a radical increase of data volume recently. The ever-increasing resource-intensive tasks, such as the high-definition video playback, and maritime navigation safety, dramatically increase the energy consumption and the requirement of bandwidth of wireless maritime networks. This Springer Brief aims at providing valuable insight on the high-efficiency data transmission scheduling, group intelligent search and rescue approach aided by the artificial intelligence (AI)-powered maritime networks. The great challenges and opportunities of the 5G networks and AI spanning from the ground to the sea are also discussed.

In Chap. 1, we present a brief introduction of maritime heterogeneous networks, including space/air/sea/ground-integrated network architecture, framework, operations, etc. In particular, the software defined network (SDN)-based maritime heterogeneous network framework is designed. In Chap. 2, a comprehensive survey is provided to cover the existing research and development in the considered field, such as maritime networking, resource allocation methods, and mission critical maritime applications. In Chap. 3, an intelligent transmission scheduling method using the deep reinforcement learning for SDN-based maritime communication network is proposed with multiple network objectives (i.e., minimizing delay, cost, or energy consumption). In Chap. 4, mobile edge computing (MEC) is introduced into maritime communication network and a multi-vessel computation offloading algorithm based on the improved Hungarian algorithm is proposed. In Chap. 5, a novel search and rescue networking is proposed, which is based on the multi-agent autonomy decision and task allocation. In this model, there are two levels of control agents, which adopt distributed edge computing to complete mission-critical applications of path search and target positioning, respectively. Conclusions are presented and open research issues are discussed in Chap. 6.

The authors would like to thank Hailong Feng, Jiabo Li, Zhi Jiang, Shan Gao, Xin Sun, Lingzheng Kong, Meng Qin, Jiacheng Chen and other Broadband Communications Research Group (BBCR) members at the University of Waterloo, for their contributions in the presented research works. Special thanks go to the editors at Springer for their help throughout the publication preparation process.

This work was supported in part by Natural Science Foundation of China under Grant 61771086, Liaoning Province Young Talents Foundation under Grant XLYC1807149, Dalian Outstanding Young Science and Technology Talents Foundation under Grant 2017RJ06.

Dongguan, China   Tingting Yang
Waterloo, Canada   Xuemin (Sherman) Shen

Contents

Chapter 1
Introduction

The overarching goal of maritime communication networks lies in satisfying the mission-critical applications and services at sea. Notwithstanding the significant traction gained by the intelligent marine networks services both for urgent search activities and perceptual modalities for future unmanned vessels, fundamentals of key enablers remain elusive in light of the challenges towards stringent reliability, latency and energy consumption requirements of mission applications. In this chapter, we first introduce the mission-critical applications and services at sea, as well the developing requirements. Moreover, the exacerbated challenges and research opportunities are listed in light of the multi-horizon, based on the analysis of unique characteristics and requirements especially for maritime search and rescue paradigm. Finally, we summarize our contributions related to architecture, computation-intensive tasks offloading and communication and computation resources allocation, which are of paramount importance towards mission-critical application driven intelligent maritime networks.

1.1 Mission-Critical Applications and Services at Sea

In maritime communication networks, how to share the decentralized resources between vessels has become a major problem, such as the needs to transmit to rescue coordination center (RCC) and shore based station reporting information, as well as voyage data recorder (VDR) [1] information in the course of navigation. In addition, it is also a great challenge to allocate the communication resources for providing the maritime information in a timely manner. Although some achievements have been made in extending land networks to maritime networks, there are still many unresolved problems (e.g., coverage areas and unstable links). Compared with traditional maritime services (e.g., sending and receiving e-mail), more and more ship users put forward higher communication quality requirements to suppliers, such as multimedia services. In the mean time, services in the 5G era have become more diverse, such as large-scale aircraft communications, mission-critical Internet, mobile broadband and so on. Therefore, mobile networks must improve their performance in terms of security, reliability, delay and mobility [2]. Therefore, the rapid growth in the number of maritime communication services needs to be accompanied by diversification and diversified needs. In this situation, researches on the allocation of marine network resources has led to new challenges as listed below:

- In maritime networks, the increasing demands for service and entertainment by ship users have led to a state of imbalance between supply and demand. Hence, the scarce resources of maritime networks can not fully meet the demands of communication resources.
- Due to the different needs of various services for maritime network resources, how to allocate network resources effectively has become the key problem [3].
- Different ships and services should be given diverse priorities on using network resources. At the same time, due to the conflict of interests, it is impossible to allocate network resources reasonably through user cooperation.

1.2 Challenges to Maritime Communications

Maritime communication networks have following challenges compared with the land communication networks:

1. The quality of service (QoS) requirements of users are continuously improving, and the existing communication links cannot meet the requirements. Communication quality and transmission rate will be seriously affected by nodes movement, offshore multi-path effects, channel fading and bad weather.
2. The number of service facilities (such as near-shore base stations) is much smaller than the number of ships, which can lead to information congestion. Hence, ships sail in different sea areas (such as the deep sea and near shore). Due to the influence of network coverage, any single network cannot cover all sea areas and cannot meet the communication needs of all ships.
3. Ship navigation is a moving scene. Due to the rapid change of network environment, the distribution of users density in ship networks is uneven.
4. Diversification of maritime communications requires different application services to have different quality requirements. For example, real-time information-related services in navigation channels require higher reliability and lower transmission delay, while video services and the Internet focus on bandwidth requirements.

Another challenge what maritime communications faces is network resources management mechanism [4]. Optimization of resource allocation in maritime communication systems are still in its infancy. The future communication fifth generation mobile networks (5G) [5] is being tried out and will become one of the most important information infrastructure. 5G technologies can be used in maritime communications to fulfill the requirements of maritime communication and channel environment. For example, the key technologies of 5G, millimeter wave multi-hop cooperative transmission technology and large-scale antenna array technology can be used to solve the problems of improving system capacity and communication coverage. It is still of great significance to develop an efficient integrated maritime communication networks and study the corresponding efficient resource management mechanism [6].

Maritime accidents mainly refer to the situation in which ships or other equipment are in danger or distress due to fire, collision, explosion, abandonment and so on, which need to be rescued quickly [7, 8]. Many countries have established search and rescue procedures [9]. In the search and rescue process, it is necessary to correctly configure the search and rescue mission according to the location of distress, sea conditions and other information, and start the search and rescue work as soon as possible [10]. The maritime search and rescue center should set up a search and rescue team to develop a detailed search and rescue plan and conduct a comprehensive assessment of the risk. At the same time, it arranges rescue ships, aircraft, and port tugboats to rescue trapped people promptly to make scientific decisions. However, the search and rescue process and measures only ensure the effectiveness of the salvage operation, and can not make a good decision. Because of the mobility, responsiveness, autonomy, and intelligence of the multi-agent network, intelligent decision-making can be used to avoid subjective decision-making errors, to achieve fast and accurate search and rescue. Therefore, multi-agent is essential for the maritime search and rescue system [11]. The maritime search and rescue system based on multi-agent system (MAS) has been developing continuously in recent years, which provides a theoretical basis for the allocation of maritime search and rescue tasks.

1.3 Our Contributions

Due to the shortcomings of traditional algorithms, such as repeatedly building models, adjusting parameters, optimizing single objective, etc., they can not meet the requirements of QoS and online information collection in the changing marine environment. Because of its complexity, the traditional algorithm is powerless in dealing with multi-objective QoS optimization problems (such as energy, cost, delay). First of all, to solve the communication problem of heterogeneous networks, the ship and shore-based networks are centrally dispatched and treated as virtual nodes, and a communication network architecture based on SDN is proposed. Secondly, the deep reinforcement learning method is adopted to solve the problem of scheduling optimization when the packet arrival rate is too high, so the deep reinforcement learning method is used to avoid large computing and storage space. At the same time, the application of maritime search and rescue and the computation-intensive tasks offloading are discussed. Main contributions are as follows:

1. A maritime communication network architecture integrating space, air, ground and ocean network is proposed, which takes multiple QoS as evaluation criteria, combines channel and cache state, and regards network resource scheduling strategy as a joint optimization problem.
2. In the problem of computation-intensive tasks offloading, we focus on the execution delay of computational tasks and the minimization of ship terminal energy consumption. First, we need to decide whether to offload it to edge servers. Second, we need to decide how to select the server and which server to run the computation task. The purpose is to calculate the execution delay based on a certain weight and minimize the energy consumption.
3. A search and rescue network model based on multi-agent autonomous decision-making and task assignment is proposed. The first level and the second level control agents use distributed edge computing to complete the tasks of path search and target location, respectively. In addition, after completing the target search, we use multiple agents to build a temporary communication network. At the same time, a communication forwarding strategy based on multi-agent joint network is designed, which makes full use of the load capacity of the network. The network model strategy can integrate the above search and rescue resources, independently analyze the scene situation, and finally provide a complete search and rescue process.

The rest of this monograph is organized as follows. In Chap. 2, the literature review and background information related to maritime communication search and rescue, mobile edge computing and maritime heterogeneous networks are introduced. In Chap. 3, the transmission scheduling problem in maritime communication networks is studied based on deep reinforcement learning. A multi-vessel offloading algorithm in maritime mobile edge computing network is proposed in Chap. 4. We discuss the application of maritime communication network based on software definition in maritime search and rescue in Chap. 5. Chapter 6 describes the future research direction of maritime communication network and summarizes the whole monograph.

References

1. Morsi, I., Zaghloul, M.S., Essam, N.: Future voyage data recorder based on multi-sensors and human machine interface for marine accident. ICCAS Gyeonggi-do **2010**, 1635–1638 (2010)
2. Kidston, D., Kunz, T.: Challenges and opportunities in managing maritime networks. IEEE Commun. Mag. **46**(10), 162–168 (2008). October
3. Mei, H., Wang, K., Yang, K.: Multi-layer cloud-RAN with cooperative resource allocations for low-latency computing and communication services. IEEE Access **5**, 19023–19032 (2017)
4. Xia, M., et al.: The state of the art and challenges of marine communications. Scientia Sinica **47**(6), 677 (2007)
5. Boccardi, F., Heath, R.W., Lozano, A., Marzetta, T.L., Popovski, P.: Five disruptive technology directions for 5G. IEEE Commun. Mag. **52**(2), 74–80 (2014). February
6. Bekkadal, F., Yang, K.: Novel maritime communications technologies. In: 10th Mediterranean Microwave Symposium, Guzelyurt, vol. 2010, pp. 338–341 (2010)

7. Zhao, M., Yao, X., Sun, J., Zhang, S., Bai, J.: GIS-based simulation methodology for evaluating ship encounters probability to improve maritime traffic safety. IEEE Trans. Intell. Transp. Syst. **20**(1), 323–337 (2019)
8. Park, J., Kim, J.: Predictive evaluation of ship collision risk using the concept of probability flow. IEEE J. Ocean. Eng. **42**(4), 836–845 (2017)
9. Weinberg, A., Kaminsky, Y., Keblawi, F.: A novel concept for a satellite-based maritime search and rescue system. IEEE Trans. Veh. Technol. **26**(3), 258–269 (1977)
10. Feng, C., Zheng, H., Zhou, L., Cheng, J.: Detection performance analysis of recurrence quantification analysis measures for low observable target within sea clutter under different sea conditions. IET Radar Sonar Navig. **9**(4), 447–456 (2015)
11. Lewis, F.L., Cui, B., Ma, T., Song, Y., Zhao, C.: Heterogeneous multi-agent systems: reduced-order synchronization and geometry. IEEE Trans. Autom. Control **61**(3), 1391–1396 (2016)

Chapter 2
Background and Literature Survey

Due to the rapid growth of maritime business data, data centers and backbone networks need to deal with a large number of complex network traffic. Resource-intensive tasks such as high-definition video playback, multimedia applications, and online games have been growing in recent years. These requirements not only increase the burden of bandwidth demand but also increase the energy consumption of the ship terminal network. At the same time, great changes have been raised in the characteristics and requirements of current maritime services, from the initial optimal requirements of point-to-point communication to support a variety of service quality. Exploring and deploying a maritime communication network with more omnidirectional coverage, higher reliability, higher speed, and lower cost will have a great influence on the development of the maritime industry and security in the future. Therefore, the introduction of brand new network technologies (such as software defined network, mobile edge computing, intelligence algorithms) into the field of maritime communications provides new ideas for solving challenges faced by maritime communications, but many basic research issues has not been thoroughly solved. We divide the existing works in the relevant literature into three categories of research problems.

2.1 SDN-Based Maritime Heterogeneous Networks

The communication network can be designed in the direction of supporting different and enough users. It is urgent to optimize the scheduling allocation and completion efficiency of all kinds of network resources. Industry and academia pay special attention to software-defined network (SDN) technology [1, 2]. It not only solves the problem of service mismatches such as transmission delay and bandwidth but also solves the problem of different protocol architectures and communication models in a heterogeneous network. SDN changes network routing from multi-hop to centralized

© The Author(s), under exclusive license to Springer Nature Singapore Pte Ltd. 2020
T. Yang and X. Shen, *Mission-Critical Application Driven Intelligent
Maritime Networks*, SpringerBriefs in Computer Science,
https://doi.org/10.1007/978-981-15-4412-5_2

through the separation of forwarding and control layers. The routing decision is realized by software programming, which makes full use of network resources. At the same time, network functional virtualization (NFV) [3, 4] is another important concept, which uses hardware and software decoupling and functional abstraction to make the functionality of network devices no longer dependent on specific hardware. To achieve the rapid development and deployment of new businesses, it is necessary to share resources flexibly. The combination of SDN and NFV [5] can realize ship-to-ship and ship-to-shore communication, and can directly program and control the abstract underlying communication connection, which can effectively improve the flexibility and efficiency in network management. Exploring an effective SDN control layer decision algorithm to solve the scalability optimization issue will be a new topic in the future wireless network.

The era of big data arises at the historic moment with the wide popularization of network services. This part introduces the practicability of SDN from four aspects: application expansion, wireless network architecture, network flow control, and network performance optimization. Major service providers have deployed a large geographically distributed information center to handle large amounts of data from a variety of smart devices. Data migration in different data centers is too heavy for network traffic to transfer packets from source to destination. Aiming at the optimization of infrastructure and network resources, Chaudhary et al. proposes a big data management method based on SDN, which fully adjusts the data storage unit and network bandwidth [6].

In recent years, cloud services have provided managers with flexible and low-cost information technology (IT) resources, including software and hardware in the wireless network framework. Accommodating most Internet traffic in the future will require a significant increase in cloud data center computing stacks and storage. However, with the proliferation of mobile devices, broadband will be under unbearable pressure when large amounts of data are wirelessly in and out of the cloud center, which can also lead to unpredictable delays. The concept of moving edges and fog [7] computation is gradually entering the field of vision. To provide location-aware and low-latency services, fog computing is required to extend the cloud to the edge of the network. Yang et al. used fog computing as an ideal complement to cloud-fog computing rather than a replacement. At the same time, to optimize the utilization of network resources and improve the quality of experience, a cloud interoperability framework supporting SDN is also created [8]. After seeking to promote more effective results, Zhang et al. proposed an intelligent cloud service management framework based on deep reinforcement learning [9]. Similarly, Cheng et al. designs a two-stage RP-TS (Resource provision-Task scheduling) processor based on deep Q learning through the concept of intelligent cloud, which uses self-learning to generate the most effective long-term decision [10].

Artificial intelligence algorithm has been widely used in vehicle network in the field of extension. In order to realize the heterogeneity of network services, Zhang et al. proposed an air-space-ground network architecture based on software definition, which divides the resources of each region and simplifies the management of the network [11]. The architecture puts all available resources into a common dynamic paves and pool the way for resource sharing between different departments. In order to improve the quality of experience (QoE) of 5G vehicle networks, Huang et al. explored a method to combine SDN heterogeneous networks with in-vehicle communications [12]. Based on the information of the vehicle, the optimal correlation between the vehicles of different states is obtained, and the QoE of the system is maximized. Lai et al. aim to address the security challenges of decentralized and centralized networks in vehicle networks. A group-based 5G-VANETs secure group communication system is proposed for the group-oriented vehicle environment. To ensure the exchange of information between vehicles with confidentiality and integrity, the authors fully consider the establishment and management of distributed network security organizations and dynamic solvable issues that enable vehicle members moving between heterogeneous networks to access the network effectively and securely [13].

To provide more efficient performance, a method of network optimization is designed to eliminate broadcast discovery packets in the network process under the software-defined network architecture. Appropriate channel allocation in (IoT) transmission network is the basic guarantee of efficient transmission [14]. Tang et al. proposed an intelligent system based on deep learning [15].

To achieve efficient network traffic control, Fadlullah et al. proposed an intelligent routing method based on deep learning, and explained the future network traffic control research [16]. Huang et al. applied this method to multimedia communication and proposes a flow control architecture based on deep reinforcement learning, which makes the model-free network management in SDN more efficient [17]. Jindal et al. applied traffic management problems to in-vehicle network physical systems and uses undisplayed packets and deep learning to plan the optimal path [18]. Liu et al. used deep learning to predict the cache efficiency of cache optimization problems. Collecting data through neuron functions translates popularity predictions into multi-classification problems to make caching efficient [19]. In order to improve the network performance of the next generation vehicle, the integrated framework proposed by He et al. can realize the computing resources, dynamic choreography, and cache of the network. At the same time, it can also be applied to liquidity prediction, traffic classification and prediction of wireless sensor networks (WSNs) [20, 21].

2.2 Mobile Edge Computing

This part deals with computation offloading and scheduling problems. In other references, the problem of computation offloading except maritime scenarios has been widely discussed. Chen et al. [22] proposes a distributed computing offloading algorithm to achieve Nash equilibrium, calculates the upper bound of the convergence time, and quantifies the centralized optimal solution of its efficiency ratio. Zhang et al. [23] proposed an online calculation algorithm, it is necessary to balance the energy collection and consumption of the sensor and stabilize its queue and data at the same time, and then the utilization of the licensed spectrum needs to be optimized while maintaining the stability of the conflict rate that can be tolerated between licensed users and unlicensed sensors.

A distributed JCORAO scheme is proposed in [24], which uses two sub-algorithms to solve the optimization problem iteratively. Based on the dynamic computation offloading algorithm based on Lyapunov optimization, a low complexity online algorithm is proposed, which determines the CPU cycle frequency of mobile execution, the offloading decision, and the transmission power [25].

To reduce the computational complexity of the computational threshold, a suboptimal resource allocation algorithm is proposed based on the resource allocation of multi-user MECO systems using time division multiple access (TDMA) and orthogonal frequency division multiple access (OFDMA) [26]. The task offloading scheduling and transmission power of the MEC system with multiple independent tasks are jointly optimized by [27], which reduces the equipment energy consumption and execution delay. According to this, a low complexity suboptimal algorithm based on alternating minimization is proposed. In order to obtain the minimum energy consumption under delay constraints, [28] combined with the multiple access characteristics of 5G heterogeneous networks, an energy-efficient computation offloading (EECO) scheme is designed to optimize offloading and radio resource allocation. In [29], Le et al. considered a multi-user MECO system with a single CVM base station. Each user can decide to offload the task in order to minimize the completion time of the user task. Wang et al. develops an optimal resource allocation scheme in the MEC framework to minimize the total energy consumption of AP subject to personal computing delays [30]. The semi-closed form optimal solution is obtained by using advanced optimization technology. In order to limit the time average energy consumption, an active service framework is proposed in [31] to minimize response delay by actively streaming blocks before a request. Wang et al. [32] uses the alternating direction multiplier method (ADMM) to solve the problem effectively. Yang et al. [33] minimizes the average completion time for all users by using off-line heuristic partitioning for multi-user computation. Lee and Zhang [34] developed a distributed energy beamforming system and applied it to multiple energy transmitters by means of energy feedback measurement to charge multiple energy receivers at the same time. Ren et al. implements transparent and edge computing to build a scalable Internet of things platform [35]. Resource management of green Iot time scale and the allocation algorithm of utility optimal resource management are proposed in [36].

There are few references on resource allocation on maritime networks. With the different scenarios at sea, the exponential growth of data traffic and the diversification of terminal types, the traditional network architecture can not adapt to the complex marine environment and the future heterogeneous multi-level network. The concept of mobile edge computing (MEC) has effectively solved the challenges of high load and low delay in the development of the mobile Internet. It is considered to be one of the key technologies of the next generation network, which has been widely recognized by industry and academia. In order to reduce the energy consumption and transmission delay of network equipment, the related request data and computing tasks are offloaded to the MEC server, which enhances the user experience.

2.3 Search and Rescue Under Maritime Communications

In researches on maritime search and rescue, the maritime search and rescue information system (SARIS) was developed by BMT to accurately predict the location of persons in distress and determine search and rescue areas [37]. During the search and rescue mission, the search and rescue resources can be allocated reasonably according to the sea condition information [38]. Meteo-France of France has developed the MOTY Maritime search and Rescue Auxiliary system, which includes real-time atmospheric circulation prediction and ocean hydrodynamic model. At present, the system has been successfully applied to oil spill trajectory tracking and sea surface drift trajectory prediction. However, the system mentioned above is more suitable for providing task a allocation scheme and real-time monitoring, collecting accident scenes, etc., which has no decisive impact on the specific implementation.

2.4 Summary

This chapter has surveyed the existing literature pertaining to maritime heterogeneous network and brand new network technologies including SDN, MEC, intelligence algorithms and application of maritime search and rescue approaches. These new technologies are used to solve maritime communication problems. In order to better understand the importance of our work, we have made a wide comparison of the existing results.

References

1. Chen, T., Matinmikko, M., Chen, X., Zhou, X., Ahokangas, P.: Software defined mobile networks: concept, survey, and research directions. IEEE Commun. Mag. **53**(11), 126–133, November 2015
2. Kreutz, D., Ramos, F.M.V., Verĺssimo, P.E., Rothenberg, C.E., Azodolmolky, S., Uhlig, S.: Software-defined networking: a comprehensive survey. Proc. IEEE **103**(1), 14–76 (2015). Jan
3. Liang, C., Yu, F.R.: Wireless network virtualization: a survey, some research issues and challenges. IEEE Commun. Surv Tutor. **17**(1), 358–380, Firstquarter 2015
4. Nguyen, V., Brunstrom, A., Grinnemo, K., Taheri, J.: SDN/NFV-based mobile packet core network architectures: a survey. IEEE Commun. Surv. Tutor. **19**(3), 1567–1602, thirdquarter 2017
5. Schulz-Zander, J., Mayer, C., Ciobotaru, B., Lisicki, R., Schmid, S., Feldmann, A.: Unified programmability of virtualized network functions and software-defined wireless networks. IEEE Trans. Netw. Serv. Manage. **14**(4), 1046–1060 (2017). Dec
6. Chaudhary, R., Aujla, G.S., Kumar, N., Rodrigues, J.J.P.C.: Optimized big data management across multi-cloud data centers: software-defined-network-based analysis. IEEE Commun. Mag. **56**(2), 118–126 (2018). Feb
7. Abbas, N., Zhang, Y., Taherkordi, A., Skeie, T.: Mobile edge computing: a survey. IEEE Internet Things J. **5**(1), 450–465 (2018). Feb

8. Yang, P., Zhang, N., Bi, Y., Yu, L., Shen, X.S.: Catalyzing cloud-fog interoperation in 5G wireless networks: an SDN approach. IEEE Netw. **31**(5), 14–20 (2017)
9. Zhang, Y., Yao, J., Guan, H.: Intelligent cloud resource management with deep reinforcement learning. IEEE Cloud Comput. **4**(6), 60–69 (2017)
10. Cheng, M., Li, J., Nazarian, S.: DRL-cloud: deep reinforcement learning-based resource provisioning and task scheduling for cloud service providers. In: Proceedings of 23rd ASP-DAC Conference, Jeju, pp. 129–134 (2018)
11. Zhang, N., Zhang, S., Yang, P., Alhussein, O., Zhuang, W., Shen, X.S.: Software defined space-air-ground integrated vehicular networks: challenges and solutions. IEEE Commun. Mag. **55**(7), 101–109 (2017). July
12. Huang, W., Ding, L., Meng, D., Hwang, J., Xu, Y., Zhang, W.: QoE-based resource allocation for heterogeneous multi-radio communication in software-defined vehicle networks. IEEE Access **6**, 3387–3399 (2018)
13. Lai, C., Zhou, H., Cheng, N., Shen, X.S.: Secure group communications in vehicular networks: a software-defined network-enabled architecture and solution. IEEE Veh. Technol. Mag. **12**(4), 40–49 (2017). Dec
14. Alasadi, E., Al-Raweshidy, H.S.: SSED: servers under software-defined network architectures to eliminate discovery messages. IEEE/ACM Trans. Netw. **26**(1), 104–117 (2018). Feb
15. Tang, F., Mao, B., Fadlullah, Z.M., Kato, N.: On a novel deep-learning-based intelligent partially overlapping channel assignment in SDN-IoT. IEEE Commun. Mag. **56**(9), 80–86 (2018). Sept
16. Fadlullah, Z.M., et al.: State-of-the-art deep learning: evolving machine intelligence toward tomorrows intelligent network traffic control systems. IEEE Commun. Surv. Tutor. **19**(4), 2432–2455, Fourthquarter 2017
17. Huang, X., Yuan, T., Qiao, G., Ren, Y.: Deep reinforcement learning for multimedia traffic control in software defined networking. IEEE Netw. **32**(6), 35–41 (2018)
18. Jindal, A., Aujla, G.S., Kumar, N., Chaudhary, R., Obaidat, M.S., You, I.: SeDaTiVe: SDN-enabled deep learning architecture for network traffic control in vehicular cyber-physical systems. IEEE Netw. **32**(6), 66–73 (2018)
19. Liu, W., Zhang, J., Liang, Z., Peng, L., Cai, J.: Content popularity prediction and caching for ICN: a deep learning approach with SDN. IEEE Access **6**, 5075–5089 (2018)
20. He, Y., Zhao, N., Yin, H.: Integrated networking, caching, and computing for connected vehicles: a deep reinforcement learning approach. IEEE Trans. Veh. Technol. **67**(1), 44–55 (2018). Jan
21. He, Y., Yu, F.R., Zhao, N., Leung, V.C.M., Yin, H.: Software-defined networks with mobile edge computing and caching for smart cities: a big data deep reinforcement learning approach. IEEE Commun. Mag. **55**(12), 31–37 (2017). Dec
22. Chen, X., Jiao, L., Li, W., Fu, X.: Efficient multi-user computation offloading for mobile-edge cloud computing. IEEE/ACM Trans. Netw. **24**(5), 2795–2808 (2016). October
23. Zhang, D., Chen, Z., Awad, M.K., Zhang, N., Zhou, H., Shen, X.S.: Utility-optimal resource management and allocation algorithm for energy harvesting cognitive radio sensor networks. IEEE J. Sel. Areas Commun. **34**(12), 3552–3565 (2016). Dec
24. Zhang, J., Xia, W., Yan, F., Shen, L.: Joint computation offloading and resource allocation optimization in heterogeneous networks with mobile edge computing. IEEE Access **6**, 19324–19337 (2018)
25. Mao, Y., Zhang, J., Letaief, K.B.: Dynamic computation offloading for mobile-edge computing with energy harvesting devices. IEEE J, Sel. Areas Commun. **34**(12), 3590–3605 (2016). Dec
26. You, C., Huang, K., Chae, H., Kim, B.H.: Energy-efficient resource allocation for mobile-edge computation offloading. IEEE Trans. Wirel. Commun. **16**(3), 1397–1411 (2017). March
27. Mao, Y., Zhang, J., Letaief, K.B.: Joint task offloading scheduling and transmit power allocation for mobile-edge computing systems. In: 2017 IEEE Wireless Communications and Networking Conference (WCNC), San Francisco, CA, pp. 1–6 (2017)
28. Zhang, K., et al.: Energy-efficient offloading for mobile edge computing in 5G heterogeneous networks. IEEE Access **4**, 5896–5907 (2016)

29. Le, H.Q., Al-Shatri, H., Klein, A.: Efficient resource allocation in mobile-edge computation offloading: completion time minimization. In: IEEE International Symposium on Information Theory (ISIT), Aachen, vol. 2017, pp. 2513–2517 (2017)
30. Wang, F., Xu, J., Wang, X., Cui, S.: Joint offloading and computing optimization in wireless powered mobile-edge computing systems. IEEE Trans. Wirel. Commun. **17**(3), 1784–1797 (2018). March
31. Zhang, D., Shen, R., Ren, J., Zhang, Y.: Delay-optimal proactive service framework for block-stream as a service. IEEE Wirel. Commun. Lett. **7**(4), 598–601 (2018). Aug
32. Wang, C., Liang, C., Yu, F.R., Chen, Q., Tang, L.: Computation offloading and resource allocation in wireless cellular networks with mobile edge computing. IEEE Trans. Wirel. Commun. **16**(8), 4924–4938 (2017). Aug
33. Yang, L., Cao, J., Cheng, H., Ji, Y.: Multi-user computation partitioning for latency sensitive mobile cloud applications. IEEE Trans. Comput. **64**(8), 2253–2266 (2015). Aug
34. Lee, S., Zhang, R.: Distributed wireless power transfer with energy feedback. IEEE Trans. Signal Process. **65**(7), 1685–1699 (2017). Apr
35. Ren, J., Guo, H., Xu, C., Zhang, Y.: Serving at the edge: a scalable iot architecture based on transparent computing. IEEE Netw. **31**(5), 96–105 (2017)
36. Zhang, D., Qiao, Y., She, L., Shen, R., Ren, J., Zhang, Y.: Two time-scale resource management for green internet of things networks. IEEE Internet of Things J. **6**(1), 545–556 (2019). Feb
37. Byun, H.: A method of indirect configuration propagation with estimation of system state in networked multi-agent dynamic systems. IEEE Commun. Lett. **22**(9), 1766–1769 (2018). Sept
38. Hoai, D.K., Van Phuong, N.: Anomaly color detection on UAV images for search and rescue works. In: 2017 9th International Conference on Knowledge and Systems Engineering (KSE), Hue, pp. 287–291 (2017)

Chapter 3
Intelligent Transmission Scheduling Based on Deep Reinforcement Learning

With the increasing diversification of ship users' communication services, the QoS of data transmission has become the limitation of the development of maritime communication. The software-defined maritime communication networks are proposed to solve the problem of communication mode obstacles in heterogeneous networks. Based on this framework, we propose a transmission scheduling scheme based on improved deep Q learning algorithm which combines the deep Q network with softmax classifier (also known as S-DQN algorithm) to improve throughput, balance delay and energy consumption. First of all, the Markov decision process (MDP) is used to realize the optimal scheduling strategy. In addition, the mapping relationship between the optimal policy and the obtained information is established by using the deep Q network in the system. When the input data arrives, after the amounts of data self-learning, the optimal strategy is made as quickly and accurately as possible. The simulation results show that the scheme is better than other traditional schemes under the different quality of service, which verifies the effectiveness of the scheme.

3.1 Software-Defined Maritime Communication Networks

Because the complex maritime communication environment has formed a coexisting heterogeneous network, a variety of maritime services can not access the network independently. The traditional heterogeneous network can not forward and control closed nodes, nor can it realize the intercommunication between networks. Therefore, we study the current deployment of offshore wireless networks and build a software-defined communication network framework that integrates space and sea, as shown in Fig. 3.1. According to the needs of users, the framework dynamically schedules the available network resources and integrates the existing heterogeneous networks to implement centralized control, which can effectively improve the efficiency of network management and simplify the packet core network.

T. Yang and X. Shen, *Mission-Critical Application Driven Intelligent Maritime Networks*, SpringerBriefs in Computer Science,
https://doi.org/10.1007/978-981-15-4412-5_3

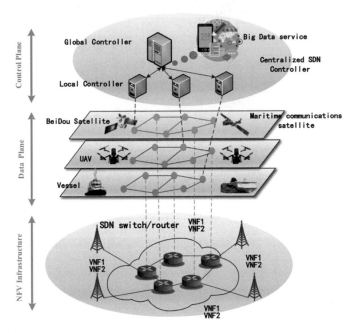

Fig. 3.1 Software-defined maritime communication networks

When a large number of big data network providers and business applications send requests to the global SDN controller, they are sent to the local SDN controller according to different service requests. Then, the local controller schedules the node uniformly through network virtualization and forwards the information back to the node in the form of a data packet. We deploy satellites, base stations, ships and drones on the line as nodes to form a loop, in which network nodes are designed to create more connections, which will greatly improve efficiency and save transmission costs. The architecture has the following technical characteristics:

Open network access platform We provide open interfaces to allow access to different networks and operators while coordinating the work of heterogeneous networks and jointly hosting wireless communication services. In addition, it can be used as the basic network of software-defined marine communication networks and is widely used in shore-based mobile communication, marine satellite communication, and 5G network.

Separate control plane from data plane The central controller with the characteristics of programmable and logical centralization can dynamically adjust the network configuration according to the demand and separate the data forwarding and control functions. Unlike infrastructure networks, which only provide a two-way data transmission mode, the central controller enables related functions such as routing and link selection.

Dynamically allocate network resources To find the best resource scheduling scheme, the software redefines the real-time operation status and business requirements of the monitoring infrastructure network of the maritime communication network. The loss rate, delay, jitter, and actual bearing capacity are judged to ensure the optimal quality of network service.

3.1.1 Channel State Model

Wireless channel varies in real time between ships, so it is necessary to model it as finite state Markov channel (FSMC) [1, 2]. FSMC model is an effective method to describe the fading process with related structural characteristics and is widely accepted.

The transition state of the channel will always occur between two adjacent states if the state of each frame is constant. We use $m \in \{1, \ldots, M\}$ to represent the Rayleigh channel with additive white Gaussian noise (AWGN). The quality of the channel is represented by the received signal-to-noise ratio (SNR). Because of the different threshold of different modulation modes, the channel state can be divided into several states, which are defined as $s_{ch}^m = \{s_0^m, s_1^m, \ldots, s_{k-1}^m\}$. The SNR is divided into $0 = \Gamma_0 < \Gamma_1 < \cdots < \Gamma_k = \infty$ by threshold values. The SNR ratio received by channel m is represented by $\Gamma\left(s_{ch}^m\right)$, when $\Gamma_k < \Gamma\left(s_{ch}^m\right) \le \Gamma_{k+1}$, then the channel m state is s_k^m. The probability that the channel state s_k^m occurs is channel state is

$$
p_{S_{ch}}\left(s_k^m\right) = \int_{\Gamma_k}^{\Gamma_{k+1}} f_s\left(\Gamma\right) d\Gamma \tag{3.1}
$$
$$
= F_s\left(\Gamma_{k+1}\right) - F_s\left(\Gamma_k\right).
$$

The cumulative distribution and probability density function of the received SNR value Γ_k are represented by the functions $f_s\left(\Gamma\right)$ and $F_s\left(\Gamma_k\right)$. Respectively, $f_s\left(\Gamma\right)$ is exponentially distributed and can be expressed as

$$
f_s\left(\Gamma\right) = \frac{1}{\bar{\Gamma}} \exp\left(-\frac{\Gamma}{\bar{\Gamma}}\right), \Gamma \ge 0, \tag{3.2}
$$

Among them, $\bar{\Gamma} = \mathrm{E}\left(\Gamma\right)$ is the average received signal-to-noise ratio. The level crossing rate at threshold $\Gamma_k, k = 1, 2, \ldots, K - 1$ is obtained by:

$$
N_{\Gamma_k} = \sqrt{\frac{2\pi \Gamma_k}{\bar{\Gamma}}} f_m \exp\left(-\frac{\Gamma_k}{\bar{\Gamma}}\right). \tag{3.3}
$$

The maximum Doppler shift is expressed as f_m. The state transition probability of channel m can be expressed as can be expressed as if each channel state is independent of each other.

$$p_{S_{ch}^m} \left(s_k^m, s_{k+1}^m \right) = \frac{N_{\Gamma_{k+1}} T_f}{p_{S_{ch}^m} (s_k)}, k \in \{0, 1, \ldots, K-1\}. \tag{3.4}$$

The combined state of M channels is $S_{ch} = \{s^1, \ldots, s^M\}$, so we express the transition probability of the combined channel as

$$p_{S_{ch}} \left(s^1, s^M \right) = \prod_{m=1}^{M} p_{S_{ch}^m} \left(s_k^m, s_{k+1}^m \right). \tag{3.5}$$

3.1.2 Cache State Model

We set the amount of incoming packets cached per unit time (1 frame) to h_{in}, and the amount of outgoing packets to h_{out}. The arrival rate is λ and the packet is subject to Poisson distribution:

$$P_{Poisson} \left(h_{in,out} \right) = \frac{\left(\lambda T_f \right)^{h_{in,out}}}{h_{in,out}!} \exp \left(-\lambda T_f \right). \tag{3.6}$$

The state of the cached node n is represented as $s_{cache}^n = \{s_0^n, s_1^n, \ldots, s_L^n\}$. h_i^n is the number of data groups in the node n cache in time slot i is. At the beginning of the next time slot $i + 1$, the amount of data in the cache is

$$h_{i+1}^n = \min \left\{ h_i^n - h_{out,i}^n + h_{in,i}^n, L \right\}. \tag{3.7}$$

The criteria for setting cache state by the system are as follows:

$$S_{cache} (h) = \begin{cases} 1, 0 < h < \frac{2}{3}L \\ \frac{1}{2}, \frac{2}{3}L \leq h \leq L \\ 0, h > L \end{cases}. \tag{3.8}$$

When $0 < h < \frac{2}{3}L$, $S_{cache} (h) = 1$ indicates that the cache state is good and allows the incoming of packet; when $\frac{2}{3}L \leq h \leq L$, $S_{cache} (h) = \frac{1}{2}$ indicates that the cache state is blocked, but data can be passed in; when $h > L$, $S_{cache} (h) = 0$ shows that the cache is full and the incoming data cannot continue. It can be seen from the above formula (3.7), (3.8) that the state transition probability of the node n corresponding to the cache in the slot i is $p_{S_{cache}^n} \left(S_{cache} (h_i^n), S_{cache} (h_{i+1}^n) \right)$.

3.1.3 Energy Consumption Model

The quadrature amplitude modulation based on η-element (η-QAM) [3, 4]. The modulation scheme of data transmission can be expressed as $U = \{u_0, u_1, \ldots, u_{u-1}\}$, $u \in [1, \infty)$, $u_0 = 2^0$ and $u_{u-1} = 2^{\eta-1}$ stand for no data transmission and ($\eta - 1$)-element phase shift keying respectively. The bit error rate of η-QAM affected by additive white Gaussian noise is approximately expressed as:

$$A_{\eta-QAM} \approx 0.2 \exp\left[\frac{-1.6 \cdot \Gamma\left(s_{ch}\right)}{(2^\eta - 1)}\right], \tag{3.9}$$

when $\eta \geq 2$ and $A_{\eta-QAM} \leq 10^{-3}$, estimation error needs to be controlled within 1dB. The packet loss rate increases with the increase of BER.

We set that the number of bits per packet remains constant, and the order of η is proportional to the amount of data transmitted in each frame and inversely proportional to the energy loss. Therefore, the efficient transmission can be transformed into the problem of minimizing energy consumption. According to the optimal decision π, node n allocates $[h_{out}]^M$ channels when sending M packets to the cache to minimize energy consumption. So the minimum power can be expressed as

$$P_w\left(s_{ch}, u_u\right) = \sum_{m=1}^{M} P\left(s_{ch}^m, u_u^{m*}\right), \tag{3.10}$$

where m^* is the channel selected by the optimal decision π. When the modulation order number is $\eta = 1$, the minimum power is expressed as

$$P_w\left(s_{ch}^m, u_0\right) \leq \frac{erfc\left(2A_{\eta-QAM}\right)^{-2} \cdot W_N}{\Gamma\left(s_{ch}^m\right)}, \tag{3.11}$$

where function $erfc\,()$ presents error function, and W_N is noise power. When the modulation order number is $\eta > 1$, the minimun power is rewritten as

$$P_w\left(s_{ch}^m, u_{u-1}\right) \leq \frac{-W_N u_{u-1} \log\left(5A_{\eta-QAM}\right)}{1.6\Gamma\left(s_{ch}^m\right)}. \tag{3.12}$$

3.2 Markov Decision Processes

This section mainly solves the problem of effective transmission of uplink data in the architecture. In the Markov decision process (MDP), the next state of the system is independent of the previous states but only related to the current state. Because this decision is more suitable for mobile scenarios, modeling it as MDP, will make online dynamic gauges effective [5, 6].

In this system, the number of data transmission is set to N, and the number of wireless channels is M. The state of the ship system and the needs of the ship users are sampled according to the basic unit time slot T_f. When the rate v of the transmitted data is constant, the node and the spectrum k remain independent of each other at the same time. Try to keep the transmission in a good channel state to maintain effective transmission. Set the maximum capacity of the system cache to L, and the capacity h will be increased by 1 when there is a packet entry. When $h > L$, he imported data will result in the loss of packets.

MDPs can be represented by $\left(S_{sys}, A', P_C, R_{total}, \gamma\right)$. The network status and demand of ship users are collected in terms of time T_f. Let S_{sys} be the system state, and all actionable sets from the current reliable channel to the next time slot are represented as A'. At the same time, the probability of the action a_i $\left(a_i \in A'\right)$ in the decision slot i in the state s_i $\left(s_i \in S_{sys}\right)$ is expressed as P_C. The return obtained by the action a_i of the decision slot i in state s_i is expressed as the reward function R_{total}, so that γ is the discount factor to satisfy $\gamma \in [0, 1)$.

Using matrix B $(M \times N)$ to represent channel allocation combination scheme

$$B = \begin{pmatrix} b_{11} & \cdots & b_{1n} \\ \vdots & \ddots & \vdots \\ b_{m1} & \cdots & b_{mn} \end{pmatrix}. \tag{3.13}$$

Among them, b_{mn} is the distribution relationship between ship n and channel m. If data packet is transmitted over channel m to the nth ship, then $b_{mn} = 1$, otherwise $b_{mn} = 0$. At the next slot, the channel allocation scheme is only related to the current state and has nonbusiness with the historical states. Meanwhile, there are two constraints as below:

$$\sum_{m=1}^{M} b_{mn} \leq 1, \sum_{n=1}^{N} b_{mn} \leq N. \tag{3.14}$$

The former shows that when the data is transferred to the cache of the corresponding ship, at most one channel allows the data to pass through. The latter shows that when the channel is allocated in a combined way, in order to avoid loop fault data packets can not be transferred to the user as a relay station.

We need to solve typical NP-hard problems with high complexity. Set the number and status of system actions to $S_{sys} = (h + 1)^N k^M$ and $A' = NMU$ respectively. Therefore, the complexity of the strategy developed by the agent is $O\left(S_{sys} A'\right) = O\left((h + 1)^N k^M\right)$.

3.2.1 System State Transition

System state $s_{sys} = \{s_{ch}, s_{cache}\}$ takes full account of cache and channel state. In time slot i, the action taken by the channel allocation scheme can be represented as $A_i = \left\{ a_{i,1}^{cache} \cdots a_{i,N}^{cache}, a_{i,1}^{ch} \cdots a_{i,M}^{ch}, a_{i,1}^{mod} \cdots a_{i,U}^{mod} \right\}$. When $a_{i,N}^{cache} = 1$, it is shown that data packet in time slot i is dispatched to the relay ship N. while $a_{i,N}^{cache} = 0$ means that no action is taken. Both $a_{i,M}^{ch} = 0\backslash 1$ and $a_{i,U}^{mod} = 0\backslash 1$ are similar to $a_{i,N}^{cache} = 0\backslash 1$. We can assume that the state transition probability of the system is independently equal and unaffected by interference. There are

$$p_c \left(s_{sys}' \right) = \frac{1}{\delta}, \quad \sum_{s_{sys}' \in S_{sys}} P_c \left(s_{sys}' \right) = 1. \tag{3.15}$$

δ is the number of elements in system state set S_{sys} and s_{sys}' represents system state in any time slot. Therefore, the state transition probability of the system can be expressed as

$$
\begin{aligned}
&P_C \left(s_{sys}' \right) \\
&= p_c \left(s_{sys}' \,|\, s_{sys}, a \right) \\
&= p_c \left(\left(s_{ch}', s_{cache}' \right) \,|\, (s_{ch}, s_{cache}), a \right) \\
&= \prod_{m=1}^{M} p_{s_{ch}} \left(s_{ch}' \,|\, s_{ch}, a \right) \prod_{n=1}^{N} p_{s_{cache}} \left(s_{cache}' \,|\, s_{cache}, a \right).
\end{aligned} \tag{3.16}
$$

3.2.2 System Reward Function

The evaluation function of the current system state transferred to other states according to a certain strategy can be represented by the system reward function $R \left(s_{sys} \right)$. The system will evaluate the optimization of channel state s_{ch}, cache state s_{cache} and minimizing energy consumption P_w. It consists of the energy consumption value, the SNR $\Gamma \left(s_{ch} \right)$ of the wireless channel and the cache state value of the terminal. $R_{once,i}$ refers to the reward obtained when the system executes action a_i of the s_i in time slot i, and the total income of the system increases with the increase of the reward function. Therefore, the one-time reward function is defined as

$$
\begin{aligned}
&R_{once} \left(i \right) \\
&= \sum_{m=1}^{M} R_{ch}^m \left(i \right) + \sum_{n=1}^{N} R_{cache}^n \left(i \right) + \sum_{m=1}^{M} R_{energy}^m \left(i \right) \\
&= \sum_{m=1}^{M} \Gamma_{s_{ch}^m} \left(i \right) + \sum_{n=1}^{N} h_n \left(i \right) - \sum_{m=1}^{M} P_w^{s_{ch}^m, u_u^{m*}} \left(i \right).
\end{aligned} \tag{3.17}
$$

Here, $R_{ch}^m(i)$, $R_{cache}^n(i)$, and $R_{energy}^m(i)$ respectively represents the feedbacks which can be acquired at slot time i in channel state m, cache state n and the energy consumption from channel m. $u_{u^*}^m$ is the modulation mode u^* selected by channel m according to action a_i. Our target is to maximize the total reward of the system, so the cumulative reward can be expressed as

$$Objective: R_{total} = \max \ \mathrm{E}\left[\sum_{i=1}^{I} \gamma^i R_{once}\left(s_{sys,i}\right)\right], \tag{3.18}$$
$$subject\ to: (11)\ (12)$$

where γ^i is the discount factor, $i \in [0, \infty)$. When i tends to infinity, γ^i tends to 0, in other words, the longer the time takes, the smaller the reward value is.

3.3 Software Defined Network Deep Q-Learning Algorithm for Data Transmission Scheduling

In the SDN controller, in order to achieve the efficiency of data transmission, the deep Q-learning algorithm is used instead of the traditional algorithm. IN this section, a new S-DQN algorithm is proposed for data transmission scheduling which is shown in Fig. 3.2. First of all, MDPs will enable the system to obtain the optimal policy, and then apply the deep Q network to establish the mapping relationship between the optimal policy and obtaining information. The system will continue to train in a self-learning manner. In order to ensure the realtime communication when the data is transmitted again, the system needs to make the best choice as soon as possible.

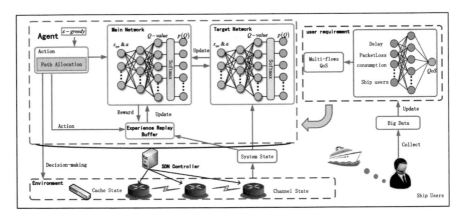

Fig. 3.2 The procedure of date transmission scheduling based on S-DQN algorithms

3.3.1 Optimal Channel Allocation Decision Based on MDPs

The modulation method is used to complete the optimal strategy of data transmission through ships and channels. MDPS system is proposed to calculate the maximum reward strategy. We defined policy π as $\pi\left(a \mid s_{sys}\right) = p\left(a, s_{sys}\right)$, which indicates the probability of selecting action a when the system is in state s_{sys}. The state value function is the current state evaluated by the system after executing π, and policy π can be expressed as

$$V^{\pi}\left(s_{sys}\right) = \mathrm{E}\left[R_{once}\left(s_{sys,0}\right) + \sum_{i=1}^{I} \gamma^{i} R_{once}\left(s_{sys,i}\right)\right]. \tag{3.19}$$

The action value function $Q^{\pi}\left(s_{sys}, a\right)$, also referred to as Q-value function, is the expected return of the system after taking action a from the state s_{sys}. So the state value function is transformed into the Behrmann equation to

$$
\begin{aligned}
V^{\pi}\left(s_{sys}\right) \\
= \sum_{a \in A'} \pi\left(a \mid s_{sys}\right) Q^{\pi}\left(s_{sys}, a\right) \\
= \sum_{a \in A'} \pi\left(a \mid s_{sys}\right)\left(R_{once} + \gamma \sum_{s'_{sys}} P_c\left(s'_{sys}\right) V^{\pi}\left(s'_{sys}\right)\right).
\end{aligned} \tag{3.20}
$$

According to π^{*}, the expected return is the largest after each decision policy action, which fully ensures the optimal strategy π of channel allocation. The optimal value function π^{*} is expressed as follows:

$$
\begin{aligned}
V^{*}\left(s_{sys}\right) &= \max_{\pi} V^{\pi}\left(s_{sys}\right) \\
&= \max_{\pi} \sum_{a \in A'} \pi\left(a \mid s_{sys}\right) Q^{\pi}\left(s_{sys}, a\right) \\
&= \max_{a} Q^{*}\left(s_{sys}, a\right).
\end{aligned} \tag{3.21}
$$

Therefore, the channel allocation problem can be transformed into an optimal solution problem according to Eq. (3.21).

3.3.2 Improved Deep Q-Learning

The current AI algorithm has achieved remarkable results in dealing with high complexity autonomous learning problems [7]. The deep Q-network (DQN) algorithm [8] particularly prominent in computer games with complex rules. Inspired by these, this chapter applies the intelligent decision algorithm of SDN controller to the complex maritime communication environment.

Deep Q-network algorithm realizes end-to-end learning from perception to action by combining reinforcement with deep learning. In recent years, researchers have optimized random sampling, network structure, Q-value and put forward an improved version [9, 10]. The new intelligent algorithm of S-DQN (the improved version of DQN [11]) solves the problem of large maritime communication space. It has the following technical characteristics:

1. The generation of loss function is a method of constructing deep learning by using the target value function of Q-learning.
2. The problem of non-static distribution and data correlation is completely solved by using empirical playback mechanism.
3. The dual network structure of primary and secondary network is established. In order to maintain the convergence of the action value function, it is necessary to use the nonlinear function approximator.
4. In order to reduce the dimension and over-fitting disaster, the deep Q network and softmax multivariate classifier [12] are combined to distribute the Q-value evenly in the quantitative interval.

The S-DQN algorithm is shown in Algorithm 1. This algorithm converts the updating problem of Q-table to a function fitting Q-value problem with θ of neural network, which is expressed as

$$\widehat{Q}\left(s_{sys}, a; \theta\right) \approx Q^{\pi}\left(s_{sys}, a\right). \tag{3.22}$$

The Q-valued function is updated according to the model-free $\varepsilon - greedy$ policy, which can be expressed in the system as

$$Q^*\left(s_{sys}, a\right) = Q\left(s_{sys}, a\right)$$
$$+ \alpha\left[R_{once} + \gamma \max_{a' \in A'} Q\left(s'_{sys}, a'\right) - Q\left(s_{sys}, a\right)\right]. \tag{3.23}$$

In the formula, α $(0 < \alpha \le 1)$ is expressed as a learning factor.

Two neural networks with the same structure are established to make the algorithm more stable. The value of target Q can be obtained through a relatively fixed parameter, called the target network. The other one is called main-net. The neural network parameter θ is continuously updated by the minimum loss function. The expression of loss function is

$$\begin{cases} L\left(\theta\right) = \mathrm{E}\left[\left(Q_{t \arg et} - Q_{main}\left(s_{sys}, a; \theta\right)\right)^2\right] \\ Q_{t \arg et} = R_{once} + \gamma \max_{a'} Q_{main}\left(s'_{sys}, a'; \theta^*\right), \end{cases} \tag{3.24}$$

The target value function is represented as the parameter θ^*. The gradient of the loss function is calculated by the following parameters:

$$\frac{dL(\theta)}{d\theta} = E\left[Q_{t\,arget} - Q(s_{sys}, a; \theta) \frac{dQ(s_{sys}, a; \theta)}{d\theta}\right]. \tag{3.25}$$

In order to achieve the end-to-end optimization goal, the stochastic gradient descent method (SGD) is used.

At the same time, deep Q-learning saves the samples produced by the interaction between agent and environment to form an empirical playback mechanism, which greatly improves the correlation between training samples. In the experience pool, each sample is a 4-tuple $e_i = (s_{sys,i}, a_i, r_{once,i}, s_{sys,i+1})$. Wherein, e_i is the experience of the ith iteration and $s_{sys,i+1} = (s_{sys,i}, a_i, x_{i+1})$. x_{i+1} represents the initial observation state in slot $i + 1$.

Assuming that the number of iterations satisfies $i = 1, 2, \ldots, I$, its corresponding set: $D = [e_1, e_2, \ldots, e_N]$. The value function is modified to

$$Q(s_{sys}, a, \theta) \xrightarrow{modification} Q(\varphi(s_{sys}), a, \theta'). \tag{3.26}$$

where $\varphi(s_{sys})$ is the state input under deep-learning. θ' includes Q-learning parameters and deep learning parameters. At this point, the experience playback mechanism is modified accordingly to

$$\begin{cases} \overline{D} = [\overline{e}_1, \overline{e}_2, \ldots, \overline{e}_I] \\ \overline{e}_i = (\varphi(s_{sys,i}), a_i, r_{once,i}, \varphi(s_{sys,i+1})) \end{cases}. \tag{3.27}$$

The S-DQN algorithm fuses softmax multivariate classifier with deep Q network. Assuming that the state s_{sys} and action a are the input, the Q-value is used as the output to generate a mapping relation to achieve fast convergence. At the same time, the output Q-value is classified twice by using softmax classifier.

$$p(Q(k)) = \frac{\exp(Q(k))}{\sum_{j=1}^{K} \exp(Q(j))}, \tag{3.28}$$

here, the probability of the kth output of Q-value is $p(Q(k))$.

After the softmax controller classifies the Q-value, the Q-value is mapped to the interval (0, 1) and normalized. This method has the following advantages: (1) softmax can intuitively see the probability of action and can give the best action corresponding to Q-value. (2) if only maximizing the Q-value to find the optimal strategy is easy to cause over-fitting, but normalizing the Q-value can solve a similar problem. (3) when the scale of the system is large, there will be a large action space, which will lead to a sharp decrease in the convergence rate of Q-value, and finally lead to the disaster of dimension. In this section, the softmax classifier is used to reduce the convergence workload of Q-value through probability calculation, so that the convergence of Q-value is not affected by the decrease, but becomes faster.

Algorithm 1: A wireless network transmission scheduling scheme based on S-DQN algorithm.

Initialize the experience replay buffer D.
Initialize the main deep-Q network with random weight θ.
Initialize the target deep-Q network with weights $\theta^* = \theta$.
for episode $= 1, \ldots, M$ **do**
    Initialize the observed value $x_1 = \{s_{ch,1}, s_{cache,1}\}$ where $s_{cache} = 0$ and s_{ch} is randomly generated and preprocess $s_{sys,1} = x_1$.
    Initialize action $a_0\,(n, m, u)$, where n, m and u are selected randomly.
    for $i = 1, \ldots, I$ **do**
        With probability ε select random action a_i.
        otherwise select $a_i = \arg\max_a Q\,(s, a; \theta)$.
        Execute action a_i in the system, obtain the reward r_i and the next observation x_{i+1}.
        Process x_{t+1} to be the next state s_{t+1}.
        Store the experience $e_i = (s_{sys,i}, a_i, r_i, s_{sys,i+1})$ into D.
        Get minibatch $e_j = (s_{sys,j}, a_j, r_j, s_{sys,j+1})$ of samples from D.
        if $t = j + 1$ **then**
            $Q_{t\,\arg et, j} = r_j$
        else
            $Q_{t\,\arg et, j} = r_j + \gamma \max_{a'} Q\left(s'_{sys}, a'; \theta^*\right)$
        end if
        Update the main deep Q-network by minimizing the Loss Function $L\,(\theta)$.
        $L\,(\theta) = \mathrm{E}\left[\left(Q_{t\,\arg et} - Q_{main}\,(s_{sys}, a; \theta)\right)^2\right]$
        and perform a stochastic gradient descent step on $L\,(\theta)$ with respect to θ.
        Every 50 steps, update the target deep Q-network parameters.
    end for
end for

3.4 Simulations of S-DQN Algorithm

In this section, the effectiveness of the proposed scheme is verified by computer simulation experiments. The simulation is divided into the following three parts. First, the channel and cache state transitions are analyzed, and the simulation parameters are set as shown in Table 3.1. Simulate the scene in MATLAB. Because the density of ocean-going ships is far less than the density of vehicles on land, five ships are regarded as relay nodes between destination and source. Then the S-DQN algorithm is compared with the original DQN algorithm and the Dijkstra algorithm in the SDN controller. Finally, the QoS performance parameters, such as delay, throughput, packet loss rate and cost, which can better reflect the effectiveness of maritime communication, are selected to discuss and compare the simulation results.

Table 3.1 Simulations parameter values

Description	Parameter value
Service arrival rate	$r = 0.1 : 0.2 : 2$
Discount rate	$\gamma = 0.9$
Negative gradient learning rate	$lr = 0.1$
Exploration coefficient	$epsilon = 0.9$
Number of slots	$i = 1 \times 10^3$
Frame length/s	$T_f = 2 \times 10^{-3}$
Doppler shift/Hz	$f_d = 50$
SNR Threshold/dB	$SNR = [1.28, 3.28, 6.28]$
Bit error ratio	$BER = 1 \times 10^{-3}$
Noise power/W	$W_n = 1 \times 10^{-3}$
Number of neurons in each layer	$[8, 15, 15, 60]$
Basic transmission rate	$v = 200\,k$ (the symbol rate)
Number of bits per group	$b = 400$
Modulation number	$\eta = 4$
Number of nodes	$n = 5$
Maximum cache capacity	$L = 5$
Independent spectrum number	$k = 3$

3.4.1 Key Points: State Transition Process Simulation

Figures 3.3 and 3.4 show the cache state transition curves for the convergence thresholds of 0.001 and 0.0001. The convergence threshold, that is, the difference between the Q-values of two adjacent iterations, is used as the criterion to evaluate whether the Q-value converges or not. When the service arrival rate is 0.1, the former only needs about 10 frames to achieve Q convergence, while the latter needs at least 200 frames. The cache state of the first user has two fluctuations in the 100th and 800th state transitions resulting in two packet losses. When the Q-value does not converge, the cache state is reduced from 5 to 3 and the first packet loss occurs. If the same cache state is generated again, it will fluctuate. After a large number of iterations of deep learning, the fluctuation range is reduced and the packet loss rate is reduced successfully.

Figures 3.5 and 3.6 show channel state transition curves with convergence thresholds of 0.001 and 0.0001, respectively, similar to cache convergence curves. As you can see in the figure, there will never be more than two levels of transition because the channel state can only be transferred to a higher or lower state, not a sibling transition. Therefore, after many iterations, it can be seen that the channel state transition process is completely random independent of the initial value, so it is necessary to adjust the data transmission adaptively.

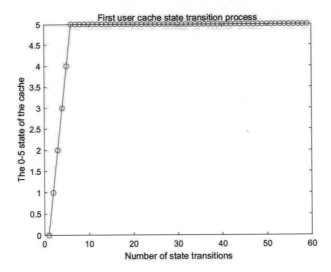

Fig. 3.3 Cache state transition curve ($conv = 0.001$)

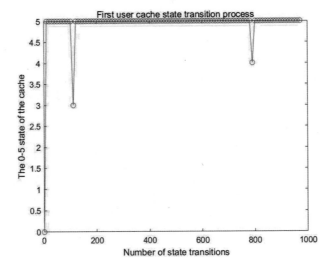

Fig. 3.4 Cache state transition curve ($conv = 0.0001$)

The channel state transition probability matrix is as follows.

$$p_{s_{ch,1}} = \begin{pmatrix} 0.2696 & 0.7304 & 0 & 0 \\ 0.4224 & 0.1888 & 0.3888 & 0 \\ 0.4424 & 0.1888 & 0.3888 & 0 \end{pmatrix} \qquad (3.29)$$

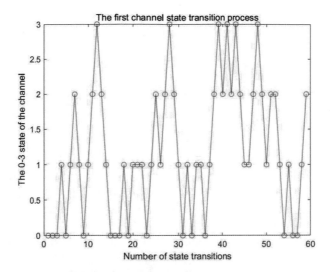

Fig. 3.5 Channel state transition curve ($conv = 0.001$)

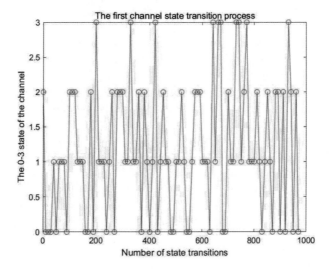

Fig. 3.6 Channel state transition curve ($conv = 0.0001$)

$$p_{s_{ch,2}} = \begin{pmatrix} 0 & 0.5302 & 0.1499 & 0.3198 \\ 0.2696 & 0.7304 & 0 & 0 \\ 0.4424 & 0.1888 & 0.3888 & 0 \end{pmatrix} \quad (3.30)$$

$$p_{s_{ch,3}} = \begin{pmatrix} 0.2696 & 0.7304 & 0 & 0 \\ 0 & 0.5302 & 0.1499 & 0.3198 \\ 0.4424 & 0.1888 & 0.3888 & 0 \end{pmatrix} \quad (3.31)$$

The randomly generated channel state is represented by the transition probability matrix and forms a joint state matrix with the node. It can be concluded from the changes mentioned above (3.29)–(3.31) that the action can determine the specific channel in which the system transmits data.

3.4.2   Algorithm Performance Simulation and Comparisons

Figures 3.7 and 3.8 present the Q-value convergence process of S-DQN algorithm after first and 1000 iterations in different initial states including $s_{sys,0}$ and $s_{sys,1}$. The simulation results show that with the increase of the number of iterations, the larger Q-value, the faster convergence speed, and the better system state.

Figure 3.9 compares the convergence speed of the S-DQN Algorithm, the original DQN Algorithm and Dijkstra's Algorithm. Dijkstra algorithm is easy to produce local optimization, which will slow down the speed of convergence. The other two algorithms accelerate the speed of convergence because they use deep learning and explore the new state. Only by quantifying the Q-value to reduce the time complexity and the amount of computation. Therefore, it has obvious advantages in the three algorithms.

Changes in the average utility and throughput of the three algorithms in a frame system are shown in Figs. 3.10 and 3.11. In the figure, utility values are used to represent the system link utilization during data transmission. The throughput of the system increases with the increase of the arrival rate of the upper packet. The cache pressure of each terminal will increase with the arrival of data, which will lead to

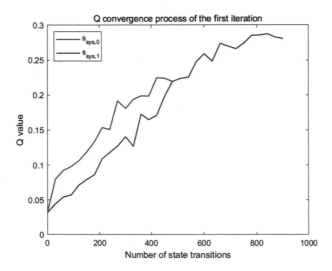

Fig. 3.7 Q-value convergence process of the first iteration (different initial values)

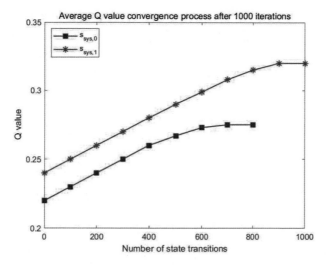

Fig. 3.8 Average Q-value convergence process after 1000 iterations (different initial values)

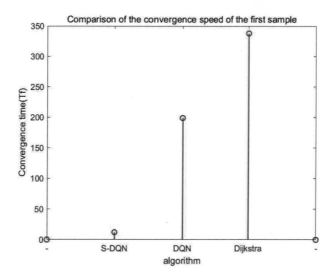

Fig. 3.9 Comparison of convergence rates of three algorithms

the reduction of the practical value of the system. In the optimal strategy, the system helps the cache reduce the pressure by increasing the output h_{out} and increasing the cost. S-DQN algorithm makes the increase of normalized Q-value will lead to the increase of packet transmission and throughput. When the packet arrival rate starts to reach 1.5, the cache pressure slowly increases. In order not to affect the transmission of system data, a higher modulation mode will be selected automatically to improve the utilization rate of the system. At the same time, in order to improve the delay

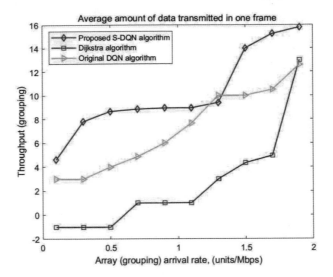

Fig. 3.10 Grouping throughput performance and comparison

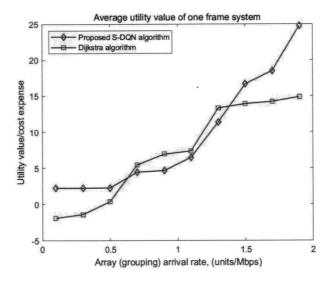

Fig. 3.11 Performance and comparison of utility value/cost expense

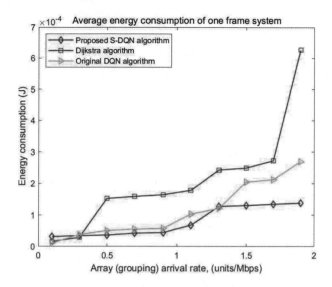

Fig. 3.12 Performance and comparison of transmission energy consumption

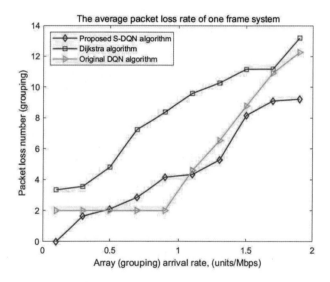

Fig. 3.13 Performance and comparison of packet loss rates

performance, the proxy increases the frequency of interaction with the environment with the change of modulation mode. The original DQN algorithm in Fig. 3.11 has relatively large utility value, so it is not compared in the same figure.

Figure 3.12 shows a comparison of the average energy consumption of the three algorithms. When the Dijkstra algorithm reaches the local optimization, the channel allocation scheme may not be optimal and the energy consumption is too large.

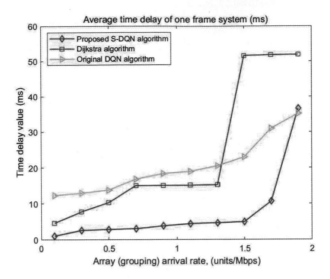

Fig. 3.14 System delay performance and comparison

Compared with the other two algorithms, the Dijkstra algorithm consumes more energy. The energy consumption curve of the S-DQN algorithm tends to be stable because of the limited cache capacity. When the amount of data reaches the maximum load limit of the cache, the cache pressure also tends to stabilize. Figure 3.13 shows that the transmission efficiency of Dijkstra's algorithm will also be reduced. Data can not be transmitted in time with the increase of service arrival rate, which leads to packet loss.

Figure 3.14 shows that the delay generated by the S-DQN algorithm is much lower than that generated by the Dijkstra algorithm. This happens because when the traffic arrival rate increases, the S-DQN algorithm makes the system cache pressure increase with the increase in throughput. In order to ensure the timeliness of the system, relay nodes are forced to choose high-order modulation mode to transmit data as soon as possible. Figure 3.15 shows that the S-DQN algorithm maintains low delay jitter. Because the S-DQN algorithm is the probability output of Q-value, it is equivalent to quantifying Q-value, so it not only reduces the fluctuation range of action but also reduces the delay jitter at the same time.

3.5 Summary

In this chapter, we have an in-depth discussion about the effectiveness of maritime communication data transmission. First of all, in order to solve the problems of communication mode in a heterogeneous network, a communication architecture based on the SDN network is proposed. Secondly, we propose a new transmission schedul-

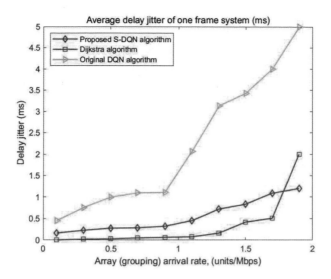

Fig. 3.15 System delay jitter performance and comparison

ing scheme. This scheme expresses the network resource scheduling problem as a joint optimization problem, considering not only the energy consumption problem but also the channel and cache state. On this basis, an enhanced deep Q-learning method is proposed to solve the over-fitting and dimension curse problems when the service data volume grows on a large scale. The simulation results explain that the scheme is superior to other traditional schemes under different service quality, and the effectiveness of the scheme is verified.

References

1. Wang, H., Yu, F.R., Zhu, L., Tang, T., Ning, B.: Finite-state Markov modeling for wireless channels in tunnel communication-based train control systems. IEEE Trans. Intell. Transport. Syst. **15**(3), 1083–1090 (2014)
2. Lin, S., et al.: Finite-state Markov modeling for high-speed railway fading channels. IEEE Antennas Wirel. Propagat. Lett. **14**, 954–957 (2015)
3. Svensson, A.: An introduction to adaptive QAM modulation schemes for known and predicted channels. Proc. IEEE **95**(12), 2322–2336 (2007)
4. Pappi, K.N., Lioumpas, A.S., Karagiannidis, G.K.: θ-QAM: a parametric quadrature amplitude modulation family and its performance in AWGN and fading channels. IEEE Trans. Commun. **58**(4), 1014–1019 (2010)
5. Li, Q., Zhao, L., Gao, J., Liang, H., Zhao, L., Tang, X.: SMDP-based coordinated virtual machine allocations in cloud-fog computing systems. IEEE Internet Things J. **5**(3), 1977–1988 (2018)
6. Li, M., Zhao, L., Liang, H.: An SMDP-based prioritized channel allocation scheme in cognitive enabled vehicular ad hoc networks. IEEE Trans. Veh. Technol. **66**(9), 7925–7933 (2017)

7. Qin, M., et al.: Learning-aided multiple time-scale SON function coordination in ultra-dense small-cell networks. IEEE Trans. Wireless Commun. **18**(4), 2080–2092 (2019)
8. Mnih, V,Kavukcuoglu, K,Silver, D., et al.: Playing Atari with deep reinforcement learning. Computer Science (2013). arXiv:1312.5602
9. Silver, D., Huang, A., et al.: Mastering the game of Go with deep neural networks and tree search. Nature **529**(7587), 484 (2016)
10. Van Hasselt, H., Guez, A., Silver, D.: Deep reinforcement learning with double q-learning. In: Proceedings of the 30th AAAI Conference on Artificial Intelligence, pp. 2094–2100 (2016)
11. Mnih, V., Kavukcuoglu, K., Silver, D., et al.: Human-level control through deep reinforcement learning. Nature **518**(7540), 529 (2015)
12. Ahmed, N.: Data-Free/Data-sparse softmax parameter estimation with structured class geometries. IEEE Signal Processing Lett. **25**(9), 1408–1412 (2018)

Chapter 4
Multi-vessel Computation Offloading in Maritime Mobile Edge Computing Network

With the rapid development of maritime communication networks, applications data of ship users has been growing exponentially, more and more resource-intensive tasks (e.g. multimedia applications, high-definition video playback, and online games) appear in daily life. New applications have enormously increased the energy consumption of ship terminals and network bandwidth requirements. To satisfy the requirements of high bandwidth and low latency for great-leap forward development and lighten the workload of mobile networks, the concept of MEC has been proposed and received extensive support from academia and industry, which is considered as one of the key technologies of next generation networks. Illuminated by this idea, this chapter presents computation offloading technology into maritime communication networks. In this chapter, we study the issue of computation offloading for computation-intensive tasks, which focuses on minimizing energy consumption of vessel terminals and time delay of computation-intensive tasks. At first, it determines whether a computation-intensive task should be offloading to a cloud server. Then, it should determine which server executes the computation-intensive task. In order to reduce execution delay and energy consumption, an improved Hungarian algorithm based on weight is proposed for multi-vessel computation offloading in maritime mobile edge computing networks. Finally, simulation results of different scenarios verify the effectiveness of the proposed algorithm.

4.1 Multi-vessel Computation Offloading

4.1.1 Computation Offloading Technology

The basic thought of MEC is to migrate cloud computing platform from inside of mobile core network to edge of the mobile access network, so as to achieve the flexible utilization of resources storage and computing. MEC systems can be

© The Author(s), under exclusive license to Springer Nature Singapore Pte Ltd. 2020   37
T. Yang and X. Shen, *Mission-Critical Application Driven Intelligent Maritime Networks*, SpringerBriefs in Computer Science,
https://doi.org/10.1007/978-981-15-4412-5_4

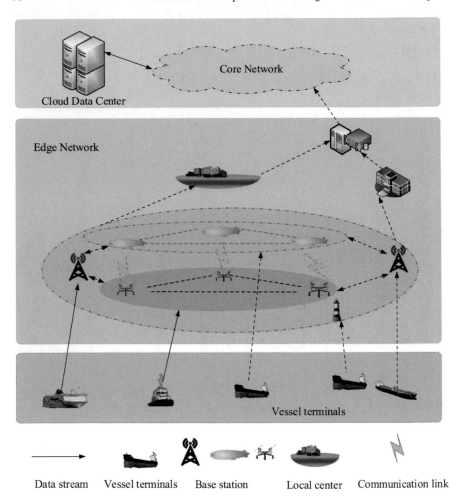

Fig. 4.1 Maritime Mobile Edge Computing Network

deployed between wireless access networks and mobile core networks in maritime communication networks. The key equipment of MEC systems is MEC servers based on the IT general hardware platform. Figure 4.1 shows a maritime communication network based on MEC and all the nodes in the block of edge network served as MEC nodes. Ship users request service applications to MEC service nodes [1]. If ship users need computing services, MEC service nodes can not provide services, then computing services of ship users will be transmitted to the core network through MEC service nodes, and service nodes of core network will provide services for ship users. If a required service is cached on MEC service nodes, which will directly serve ship users and send back the corresponding instructions.

The characteristics of MEC-based maritime communication networks are as follows:

1. *Reduce response delay*
   MEC service nodes are closer to ship users than traditional networks and greatly reduce service time and response latency.
2. *Less resource occupancy*
   A large number of popular requirements are cached at MEC service nodes, so duplicate service applications will not be transmitted to the cloud server through the core network and obtain required services on MEC service nodes directly. Therefore, MEC system reduces duplicate requests in core network and occupancy of marine network resources.
3. *Position awareness*
   MEC servers use signaling information obtained from wireless networks to determine the location of each connecting ship, which can lay a good foundation for subsequent location-based services, analysis, and other business applications.

4.1.2 Offloading Judgement

There are two standards for deciding whether to offloading computing tasks: whether to reduce latency and save energy [2].

4.1.2.1 The Execution Delay

One of the foundations of offloading decisions is to improve application performance and reduce execution delay. s_{local} is execution speed at local servers, G_r represents how much computation resources are needed to finish a computation-intensive task. If a computation-intensive task is executed locally, the time is G_r/s_{local}.

B_w means the bandwidth of the existing network, s_{edge} is execution speed at edge serves, and the input task size is $Data$. The execution time of computation-intensive task which offloads to edge serves is $\frac{Data}{B_w} + \frac{G_r}{s_{edge}}$.

We assume that the local execution time of a computation-intensive task is longer than edge execution time. The corresponding relationships are as follows:

$$\frac{G_r}{s_{local}} > \frac{Data}{B_w} + \frac{G_r}{s_{edge}}. \tag{4.1}$$

It can be concluded that

$$G_r \left(\frac{1}{s_{local}} - \frac{1}{s_{edge}} \right) > \frac{Data}{B_w}. \tag{4.2}$$

In the case of the above formula, G_r is the computing resources which should meet the computing offloading requirements as much as possible. Because the local execution speed is limited ($s_{local} < s_{edge}$) while meeting the above requirements, the local execution speed of s_{local} is as large as possible, edge execution speed is as small as possible, which realizes requirement of improving performance of vessel terminals.

4.1.2.2   The Energy Consumption

The lower the energy consumption, the higher the efficiency and the more environmentally friendly. We take this into account and offloading is more reasonable. P_G is the executing power of vessel terminals, and according to the above formulas, we get the energy consumption of local execution

$$P_G \times \frac{G_r}{s_{local}}. \tag{4.3}$$

P_1 is offloading power of vessel terminals. When a computation-intensive task is completed in edge severs, vessel terminals need to receive results of edge serves through network. The power of access connection between vessel terminals and networks is P_2, and energy consumption of offloading a computation-intensive task is formulated as follows:

$$P_1 \times \frac{Data}{B_w} + P_2 \times \frac{G_r}{s_{edge}}. \tag{4.4}$$

The energy consumption of local execution is greater than that of offloading to edge severs.

$$P_G \times \frac{G_r}{s_{local}} > P_1 \times \frac{Data}{B_w} + P_2 \times \frac{G_r}{s_{edge}}. \tag{4.5}$$

It can be concluded that

$$G_r \times \left(\frac{P_G}{s_{local}} - \frac{P_2}{s_{edge}} \right) > P_1 \times \frac{Data}{B_w} \tag{4.6}$$

4.2   Minimize Time Delay and Energy Consumption

In this section, we construct a scenario of multi-vessel and multi-MEC service nodes, which can be used for offloading and computing network resources, and make full use of maritime communication networks resources.

We use $U = \{u_1, u_2, ..., u_m\}$ to denote vessel terminals, $S = \{s_1, s_2, ..., s_i\}$ presents MEC service nodes, and $C = \{c_1, c_2, ..., c_j\}$ is the service channels set. The relationship between service channels and MEC service nodes is shown by matrix M as follows:

$$M_{c_j, s_i} = \begin{cases} 1, \text{ a task is offloaded to a MEC node} \\ 0, \text{ otherwise} \end{cases} \tag{4.7}$$

In a MEC service nodes model, the tradeoff of time delay and energy consumption is used as the total cost of connecting each vessel terminal to MEC service nodes.

4.2.1   Time Delay

When a vessel terminal offloads computation-intensive tasks to edge servers, the overhead of time delay consists of three parts, one is the time for offloading input data to edge servers, and ensure that edge servers can execute the computation-intensive task; one is the execution time of the computation-intensive task in edge servers; one is that results go back from edge servers return to vessel terminals. The overhead of time delay of vessel terminals during the entire computation offloading process can be represented as

$$T = t_{tr} + t_{exe} + t_{down}. \tag{4.8}$$

Among them, t_{up} denotes the time it takes for vessel terminals to offload input data of a computation-intensive task to edge servers, while t_{exe} is the total time which takes for the edge to execute the computational instruction to finish the execution. t_{down} is the time when results return from edge servers to vessel terminals. Since last part of backhaul time delay does not depend on characteristics of vessel terminals, it is expressed as

$$\Delta T_{u_m}^{c_j} = t_{tr} + t_{exe}. \tag{4.9}$$

The vessel terminal u_m offloads a computation-intensive task to a MEC service node s_i through channel c_j. In wireless transmission stage, a vessel terminal can select a slotted fast fading channel, offload input data to a MEC service node, and get the channel state information (CSI) [3]. The offloading rate of selected channel is

$$R = w_j \log_2 \left(1 + \frac{p_{tr} h^2}{N_0}\right). \tag{4.10}$$

where w_j is the bandwidth of channel c_j, p_{tr} represents offloading power of a vessel terminal u_m, h is the fading coefficient of channel c_j, and N_0 is the power of noise.

Moreover, $Data$ is used as the input data size of a vessel terminal to offload. Then, it can be concluded as

$$t_{tr} = \frac{Data}{w_j \log_2 \left(1 + \frac{p_{tr} h^2}{N_0}\right)}. \tag{4.11}$$

The offloading matrix M indicates whether computation-intensive tasks are offloaded to the edge servers. Based on the offloading matrix M, the number of vessel terminals on MEC service nodes can be known. r_n is the computing resource for the MEC service node s_i. We assume that a computation-intensive task is offloaded to edge server s_i, and vessel terminals share the computing resources which can be shown as

$$\bar{r} = \frac{r_n}{\sum_1^N M_{c_j, s_i}}. \tag{4.12}$$

The CPU of edge server to execute computation-intensive task is H_{ed}, and it can be concluded that:

$$t_{exe} = \frac{H_{ed}}{\bar{r}} = \frac{H_{ed}}{\frac{r_n}{\sum_1^N M_{c_j, s_i}}} = \frac{H_{ed} \sum_1^N M_{c_j, s_i}}{r_n} \tag{4.13}$$

Based on above analysis, total time delay of the whole computation offloading of vessel terminals can be shown as

$$\Delta T_{u_m}^{c_j} = \frac{Data}{w_j \log_2 \left(1 + \frac{p_{tr} h^2}{N_0}\right)} + \frac{H_{ed} \sum_1^N M_{c_j, s_i}}{r_n} \tag{4.14}$$

4.2.2 Energy Consumption

The ship terminal offloads computation-intensive tasks to edge server to consume energy. The offloading power is p_{tr} and offloading delay is t_{tr}, we have

$$E_{u_m}^{c_j} = p_{tr} t_{up}. \tag{4.15}$$

The overhead of the whole offloading and computation is

$$C_{u_m}^{c_j} = \alpha \Delta T_{u_m}^{c_j} + \beta E_{u_m}^{c_j}. \tag{4.16}$$

Among them, $C_{u_m}^{c_j}$ is the overhead of vessel terminal u_m to offload computation-intensive task through access channel c_j. The weight parameters of time delay and energy consumption are α and β respectively, and $\alpha + \beta = 1$, which are set to balance the individual requirement of the ship users. Some tasks have high requirements on latency, and some tasks have high requirements on energy consumption. For example, if a task requires high latency, then alpha will be higher than beta. If a task requires high energy consumption, then alpha will be lower than beta, but the same thing is $\alpha + \beta = 1$.

4.2.3 Optimization Target

Based on the choice of MEC service nodes, vessel terminals offload computation-intensive tasks with lower energy consumption and time delay. In order to save the overhead of vessel terminals, we can express the problem as bellow:

$$\min \sum_{m=1}^{|U|} \sum_{j=1}^{|C|} C_{u_m}^{c_j} M_{u_m}^{c_j} \tag{4.17}$$

Subject to

$$M_{u_m}^{c_j} \in \{0, 1\}, \forall c_j, u_m$$
$$S_{local} < S_{edge} \tag{4.18}$$

The maximum offloading power of the vessel terminals is $p_{\max}$, and the minimum offloading power is $p_{\min}$.

4.3 Optimal Energy Consumption Algorithm

4.3.1 Improved Hungarian Algorithm

Hungarian mathematician Edmonds proposed the Hungarian algorithm in 1965, which is the most general algorithm for two-graph matching problems. The key of the algorithm is to find path augmentation that maximizes the matching between two graphs. The model built by Hungarian algorithm which has high accuracy and could address allocation problem effectively and simply.

Classical Hungarian algorithm is an optimization method to solve the allocation problem which is based on the theorem of independent zero elements in matrix (proposed by Hungarian mathematician D. Koning). The theoretical basis of the method is that any addition or subtraction of a constant in a efficiency matrix (cost matrix) will not alter the optimal allocation schedule [4].

The basic detail steps of classical Hungarian algorithm are as below:

1. *Step A*: build the efficiency matrix $M_0(m * n)$ of resource allocation problem.
2. *Step B*: minus a minimum value from each row of the efficiency matrix M_0, and each row has a zero value and obtains M_1.
3. *Step C*: minus a minimum value from each column of efficiency matrix M_1, and each column has a zero value and obtains M_2.
4. *Step D*: M_3 is given by covering a zero value in M_2 with a least straight line. If the minimum value of lines is equal to m, go to *Step F*. Otherwise, go to *Step E*.
5. *Step E*: efficiency matrix M_3, all elements which are not covered by a straight line are subtracted from the minimum value in the not covered element, and the minimum value is added to the line intersection point to obtain M_4, then $M_2 = M_4$ is turned into *Step D*.
6. *Step F*: from the minimum row or column of zero elements, all tasks will be allocated and the optimal allocation scheme is obtained.

In above steps, assuming that $m = n$, efficiency matrix M_0 is a square matrix. In view of practice problem, the number of practice tasks is not equal to the number of workers, a common approach is to add virtual workers or tasks.

To address above problems, the concept of virtual channels is introduced to match service channels and vessel terminals. $U = \{u_1, u_2, ..., u_m\}$ is vessel terminals set, $C = \{c_1, c_2, ..., c_j\}$ is service channels set. The relationship between mobile edge computing service nodes and service channels is represented by efficiency matrix M. If the number of MEC service nodes and the number of service channels are not equal, we should add virtual service channels, and apply Hungarian algorithm to complete the efficiency matrix to decide whether the optimal matrix is satisfied until it is obtained. The algorithm is shown in Fig. 5.9 and Algorithm 1 (Fig. 4.2).

4.3.2 Optimal Energy Consumption

In the computation offloading process, we pay more attention to the energy consumption of the vessel terminals. Therefore, we consider the problem of minimizing the energy consumption of the vessel terminals.

$$E_{u_m}^{c_j} = p_{tr} t_{tr}. \tag{4.19}$$

The energy consumption of vessel terminals during offloading:

$$\min E_{u_m}^{c_j}. \tag{4.20}$$

Algorithm 1: Improved Hungarian Algorithm

1: Selection matrix
   $u_m \in U$ and $c_j \in C$
2: **for** $i \leftarrow 1$ to S **do**
3:    Compute s_i is part of MEC service nodes
      Channel c_j is part of MEC service channels
4:    **if** the result $= 1$ **then**
5:       Offloading intensive tasks to edge servers
6:    **end if**
7:    **if** the result $= 0$ **then**
8:       Locally execute
9:    **end if**
10: **end for**
11: Add virtual service channels
12: Apply the Hungarian algorithm for the efficiency matrix
13: Obtain the result efficiency matrix and decide whether it is optimal
14: **if** YES **then**
15:    OUT
16: **end if**
17: **if** NO **then**
18:    Repeat above steps
19: **end if**
20: Until optimal efficiency matrix

Subject to

$$p_{\min} \leq p_{tr} \leq p_{\max}. \tag{4.21}$$

The maximum offloading power of the vessel terminals is $p_{\max}$, and the minimum offloading power is $p_{\min}$.

We have

$$\min \frac{Data\, p_{tr}}{w_j \log_2 \left(1 + \frac{p_{tr} h^2}{N_0}\right)}. \tag{4.22}$$

The above formula $\frac{Data\, p_{tr}}{w_j \log_2 \left(1 + \frac{p_{tr} h^2}{N_0}\right)}$ is a non convex function, which is considered to be hard to address, because feasible domains sets may have a great deal of local optimal solutions, and the complexity of addressing global optimal algorithms is exponential (NP-hard). To remedy the optimization problem, we apply the objective function of non-linear fractional transformation. Therefore, the optimal value of energy consumption E^*[5] is introduced.

$$E^* = \frac{Data\, p_{tr}^*}{w_j \log_2 \left(1 + \frac{p_{tr}^* h^2}{N_0}\right)} = \min \frac{Data\, p_{tr}}{w_j \log_2 \left(1 + \frac{p_{tr} h^2}{N_0}\right)} \tag{4.23}$$

Fig. 4.2 Improved
Hungarian algorithm

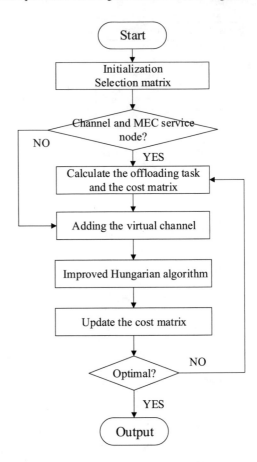

We have

$$\min Data\,p_{tr} - E^* w_j \log_2 \left(1 + \frac{p_{tr} h^2}{N_0}\right) = 0 \tag{4.24}$$

$$Data\,p_{tr}^* - E^* w_j \log_2 \left(1 + \frac{p_{tr} h^2}{N_0}\right) = 0 \tag{4.25}$$

where $Data\,p_{tr} \geq 0$, and $w_j \log_2 \left(1 + \frac{p_{tr} h^2}{N_0}\right) \geq 0$. And the objective function is
defined as

$$F(p_{tr}) = Data\,p_{tr} - E^* w_j \log_2 \left(1 + \frac{p_{tr} h^2}{N_0}\right) \tag{4.26}$$

First-order derivation, $F(p_{tr})$ on p_{tr}:

$$\frac{dF(p_{tr})}{dp_{tr}} = \frac{d\left(Data\,p_{tr} - E^*w_j\log_2\left(1 + \frac{p_{tr}h^2}{N_0}\right)\right)}{dp_{tr}} = Data - \frac{E^*w_jh^2}{Data + p_{tr}h^2} \tag{4.27}$$

Second-order derivation, $F(p_{tr})$ on p_{tr}:

$$\frac{d^2F(p_{tr})}{dp_{tr}^2} = \frac{E^*w_jh^4}{\left(Data + p_{tr}h^2\right)^2} \geq 0 \tag{4.28}$$

Evidently, the derivative of second-order is lager than 0, the energy consumption function $F(p_{tr})$ is a convex function.

Algorithm 2: Optimal Energy Consumption Algorithm

1: Input $Data$, h^2, w_j, N_0
2: Output $E_{u_m}^{cj}$, p_{tr}
3: Initialization
4: Set p_{tr}, solve $\dfrac{Dp_{tr}}{w_j\log_2\left(1+\frac{p_{tr}h^2}{N_0}\right)} = E_{u_m}^{cj}$, we have E^*
5: $p_{tr} = p_{\min}$, $E = 0$
6: **if** $Data\,p_{tr} - E^*w_j\log_2\left(1 + \frac{p_{tr}h^2}{N_0}\right) \leq E_{\max}$ **then**
7:    True, $p_{tr}^* = p_{tr}$, $\dfrac{Data\,p_{tr}}{w_j\log_2\left(1+\frac{p_{tr}h^2}{N_0}\right)} = E^*$
8: **end if**
9: Set $\dfrac{Data\,p_{tr}}{w_j\log_2\left(1+\frac{p_{tr}h^2}{N_0}\right)} = E$, increase iterations
10: Untill True or Maximum iterations
11: Optimal energy consumption

4.4 Simulations of Different Scenarios

We appraise the performance of the proposed algorithm through simulations in the MATLAB environment. Simulation parameters are set up for different scenarios.

4.4.1 Different Scenarios Comparisons

We construct a scenario of a multi-vessel and a multi-MEC service nodes model, which is used to offload intensive tasks, compute network resources, and make full use of maritime communication network resources. We assume that a service channel only serves one vessel terminal. Vessel terminals can finish a series of actions (e.g. MEC service nodes access, computation-intensive task offloading, and receive computation results) in selected time. For a given selection time, we randomly generate the number of tasks of each vessel terminal, the resources for each task, and the energy consumed by local/remote execution tasks.

We consider the selection relationship between MEC service nodes and vessel terminals. Simulation environment setting: maritime communication networks scene contains 5 base stations, and each base station contains 2 services channels, 10 vessel terminals can be offloaded at a time. Each service channel bandwidth is 5MHz, and each service channel only serves one vessel terminal within 1 selection time. And the number of vessel terminals is distributed near 5 base stations randomly. The offloading power of vessel terminals is obtained randomly from 100 to 1000 mW. And the size of computation-intensive tasks is randomly generated. Moreover, the computation resource on edge servers is 10 GHz. We also take the effects of services channel loss and noise into consideration.

Figure 4.3a presents the selection relationship between MEC service nodes and vessel terminals, which meets the above conditions. The computation-intensive task is offloaded from the vessel terminals through the service channels to the MEC service nodes. Figure 4.4a presents that time delay and energy consumption varies with the number of MEC service nodes. Time delay and energy consumption first increase as MEC service nodes grow in number. And when the number of service nodes reach 40, the total overhead peaks and then decreases. Figures 4.3b and 4.4b are the results of altering simulation parameters. In Fig. 4.4b, after altering the simulation parameters, time delay and energy consumption significantly reduced. As MEC service nodes increase, the tradeoff of time delay and energy consumption is maintained. When the point of transition is reached, the reduction in time delay and energy consumption is more pronounced and the performance has been clearly improved.

4.4.2 Performance and Comparisons

In Fig. 4.5a, we conclude that when the number of MEC service nodes and vessel terminals increases, the tradeoff of time delay and energy consumption will increase. It is sensible that with the increase of MEC service nodes and vessel terminals, more computing resources are needed.

In Fig. 4.5b, we set the weight coefficients of time delay α and energy consumption β ($\alpha + \beta = 1$), respectively. The weight of coefficients is set to balance the requirements of ship users for different types of tasks. For example, online games require lower time delay, but have higher energy consumption. Compared the results of Fig. 4.5b with Fig. 4.5a, we adjust α and β, the total overhead decreases, the strategy has lower time delay and energy consumption, which proves effectiveness of proposed algorithm.

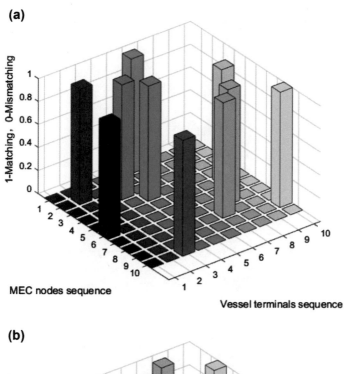

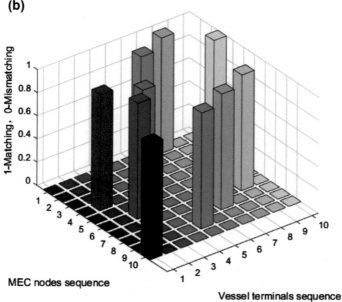

Fig. 4.3 The selection relationship between vessel terminals and MEC service nodes

Fig. 4.4 The cost for different value of MEC service nodes

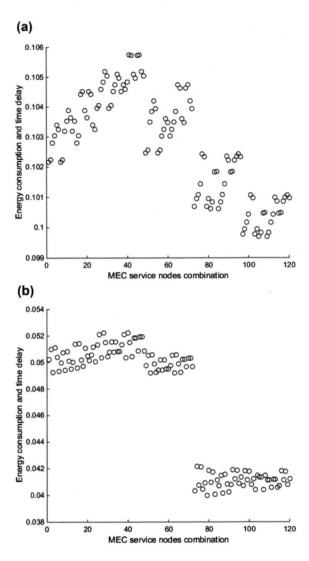

4.4.3   The Saturation Time

In Fig. 4.6a, we conclude that as the number of MEC service nodes and vessel terminals increases, the time for computing resource saturation becomes less. The proposed algorithm has a higher priority in remaining time. In this case, more intensive tasks are offloaded, which proves effectiveness of proposed algorithm. In Fig. 4.6b, performances are improved even when parameters are changed.

Fig. 4.5 The cost for different value of vessel terminals and MEC service nodes

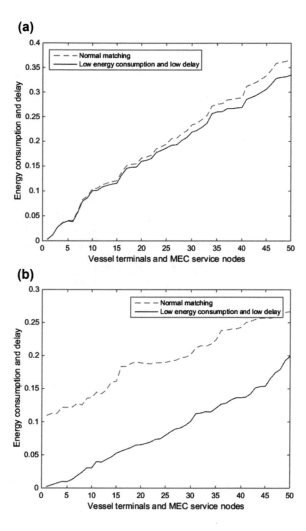

4.5 Summary

In maritime communication networks, the greatly increasing demands for ship users, including maritime affairs services and crew entertainment, which enormously exceeds the capability of the network. In order to satisfy the needs of high-speed development of mobile networks for high bandwidth and low latency, and reduce network workloads, the idea of MEC has been proposed and widely supported by academia and industry, which is considered as one of the key technologies of next generation network. And inspired by MEC, we introduce computation offloading technology into the maritime mobile cloud network. The problem of offloading computation-intensive tasks of vessel terminals is studied, energy consumption and

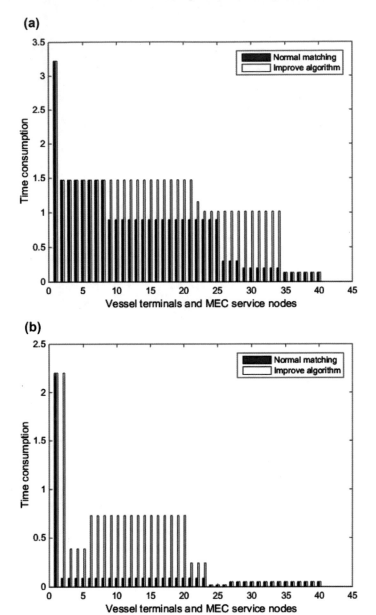

Fig. 4.6 Saturation time for different value of vessel terminals and MEC service nodes

time delay are mainly studied. Firstly, we need to decide if offload tasks to edge servers. Secondly, which server should be chosen to execute computation-intensive tasks. The purpose is to minimize the energy consumption of vessel terminals and time delay of computation-intensive tasks on a weight-based basis. In order to reduce time delay and energy consumption, this chapter proposes a multi-vessel computing offloading algorithm for maritime MEC network based on improved Hungarian algorithm. Finally, simulation results prove the effectiveness of the proposed algorithm.

References

1. Taleb, T., Dutta, S., Ksentini, A., Iqbal, M., Flinck, H.: Mobile edge computing potential in making cities smarter. IEEE Commun. Mag. **55**(3), 38–43 (2017). March
2. Mebrek, A., Merghem-Boulahia, L., Esseghir, M.: Efficient green solution for a balanced energy consumption and delay in the IoT-Fog-Cloud computing. In: IEEE 16th International Symposium on Network Computing and Applications (NCA). Cambridge, MA, USA, pp. 1–4 (2017)
3. Maddah-Ali, M.A., Tse, D.: Completely stale transmitter channel state information is still very useful. In: 2010 48th Annual Allerton Conference on Communication, Control, and Computing (Allerton), Allerton, IL, pp. 1188–1195 (2010)
4. Aslam, S., Shahid, A., Lee, K.G.: IMS: Interference minimization scheme for cognitive radio networks using Hungarian algorithm. In: The First International Conference on Future Generation Communication Technologies, London, pp. 17–21 (2012)
5. Dinkelbach, W.: On nonlinear fractional programming. Manag. Sci. **13**(7), 492–498 (1967)

Chapter 5
Mission-Critical Search and Rescue Networking Based on Multi-agent Cooperative Communication

Improving the existing maritime search and rescue network and carrying out a fast and effective search and rescue activities have become a research hotspot. Intelligent devices such as an unmanned surface vehicle (USV) and unmanned aerial vehicle (UAV) can be used to build multi-layer networks. Their fast maneuverability and distributed computing performance provide us with new research directions. In this chapter, a new search and rescue system is designed which applies the ant colony optimization (ACO) and particle swarm optimization (PSO) for search and rescue decision-making and network scheduling. Each device in the system carries out information sharing, edge calculation, and realizes autonomous synchronous search and rescue. We use ACO to plan search path and PSO to schedule data packet forwarding to build a complete maritime search communication network. This network adopts the distributed cluster control mode, reduces the calculation consumption of the control center, and improves the search and rescue efficiency. The simulation results show that the proposed maritime search and rescue solution can achieve a good performance.

5.1 Model of Multi-agent Search and Rescue

To mitigate the problem of a multi-agent search and rescue process in the actual situation, a multi-agent rescue model is established at sea. The model includes multiple elements of shore-based communication stations, communication satellite systems, central control agents, group control agents, and rescue agents. The specific model is shown in Fig. 5.1. The entire system consists of base stations, satellites, and multiple intelligent groups. The terminal decision maker sends scheduling commands to the multi-intelligent group via the base station and satellite. The multi-intelligent components are combined with three parts: central control agent, group control agent, and search and rescue agent. The central control agent is responsible for determining the

T. Yang and X. Shen, *Mission-Critical Application Driven Intelligent Maritime Networks*, SpringerBriefs in Computer Science, https://doi.org/10.1007/978-981-15-4412-5_5

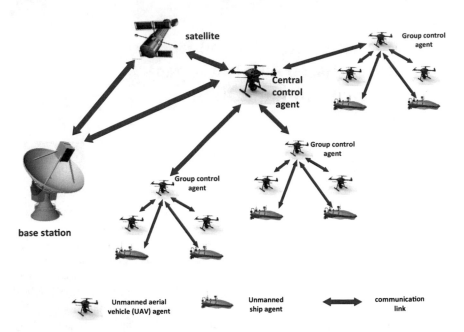

Fig. 5.1 Hierarchical maritime search and rescue model based on multi-agent (auxiliary relay communication via satellite)

location at sea area and sending search and rescue missions decisions. The group control agent is responsible for calculating the search path and searching for the shortest path [1]. Search and rescue personnel is responsible for specific search and rescue work. Where the agents of three parts complete the function of edge computing. Overall, the whole system is provided with strong autonomy and intelligence.

5.1.1   Sea Search and Rescue Process Based on Multi-agent

Combining the mature UAV, USV and other multi-agent technologies [2], can reduce personnel investment while improving search and rescue efficiency. On this basis, a novel search and rescue scheme is put forward. The flow chart of maritime search and rescue is shown in Fig. 5.2.

First, marine equipment such as people, boats or airplanes at risk emits an alarm message, after which the automatic processing system of the search and rescue center receives the alarm information and eliminates the false alarm information. The center then notifies the standby search and rescue personnel via satellite or land-based communications equipment. The information on the notice is combined with the latitude and longitude coordinated the area of the crash, sea state, weather, etc.

Fig. 5.2 Process of
maritime search and rescue
based on multi-agent

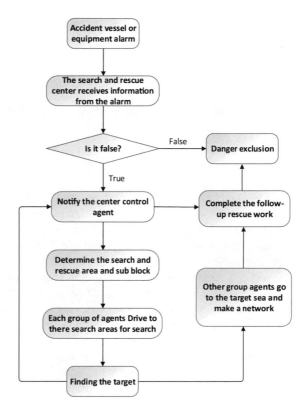

The search and rescue agents received mission and were combined with the central control agents to calculate the extent of the damaged sea area by factors such as currents and winds. After the search and rescue team receives the mission information, each group of control agents calculates the shortest path to the search and rescue area according to the sea state. After getting the optimal path, every agent group enters the search and rescue area. Until they reach the search area, every group control agent calculates the search path and sends the optimal path to all agents in the group.

After finding the search and rescue target, the search agent sends the target information to the group control agent. The group control agent then sends the information to the adjacent search group. The team towards finding the target is responsible for tracking the target and implementing appropriate rescue. Meanwhile, the group notified the surrounding large ships to provide assistance. After arriving at the crash site, the search and rescue teams establish a temporary communication network to record and transmit the situation on the ground. In this way, the land-based personnel could make decisions and then develop subsequent rescue plans [3].

In the end, the mission of search and rescue is completed. Each group uploads the information which is recorded during the search process to the central control agent for analysis and backup. Eventually, the standby state is entered.

5.1.2   Multi-agent Collaborative Networking

After completing the task of searching and positioning the shipwrecked ship, it is necessary to build a communication link between the site of the wrecked ship and the shore-based command center. The current global maritime distress and safety system mainly relies on satellite communication [4], ground radio communication, maritime safety information broadcasting and other subsystems to carry out maritime search and rescue work. Due to the limitations of the existing system infrastructure, its communication bandwidth is limited, and only a small amount of data forwarding can be completed. However, maritime distress search operations often require the support of a large number of data forwarding services, such as clear images or even video. He traditional GMDSS system does not meet the requirements of the current rescue positioning and command center real-time communication. For distributed D2D (Device-to-Device) communication [5], a temporary maritime communication network is constructed with multi-agent technology in this chapter. According to the distributed edge computing ability of multi-agent, the network quality of service (QoS) indexes, such as communication channel capacity, delay and so on, can be analyzed in real-time. Through the multi-agent network, based on a certain network communication capability, the intelligent scheduling algorithm is used to complete the data forwarding. This will take full advantage of the carrying capacity of the network to provide better and more efficient and reliable communication services for maritime rescue [6].

The Multi-agent collaborative networking model is shown in Fig. 5.3. After finding the wrecked vessel, the search agent near the wrecked location issues flood information and notifies other search nodes in the network to notify the information. At the same time, the agents near the crash site collect basic information about the crash site about the video, audio, weather, and sea conditions, etc. These agents will use their edge calculation capabilities to perform initial filtering and preprocessing of this information. Through the processing and analysis of these packets, the bandwidth pressure in the subsequent forwarding process can be reduced [7]. These agents then establish a communication network between the crash location and the shore-based base station. Each group of the agent as a communication community, the group control agent is in charge of the coordination location, and devices in the group are responsible for data forwarding. At the same time, the forwarding strategy between network nodes or cells is designed in the model to improve communication efficiency by utilizing network resources [8].

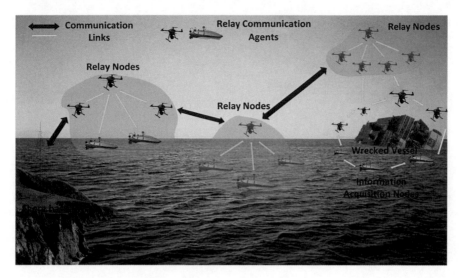

Fig. 5.3 Multi-agent collaborative networking model

5.2 Establishment of Temporary Communication Network

5.2.1 Planning of Route to Reach Search and Rescue Area

The cause of marine accidents is often bad weather conditions, which may cause the target to lose power. Therefore, the target to be estimated and located. After receiving the information about the coordinate position of the wreckage, the central control agent calculates the drift direction of the target based on the current real-time basic information and the direction of the wind. Then the search basis point is determined. The illustration about central control agent determines the search area is shown in Fig. 5.4, divides the search and rescue area, and sends it to the search and rescue agent group with the calculated reference point as the origin of coordinates, combined with the electronic chart and real-time weather information. The calculation process is as follows:

- Denote t_a and t_b as the alarm time of the ship and the starting time of the agent groups respectively, then the drift time of the wrecked ship is t.

$$t = t_b - t_a \tag{5.1}$$

- Denote v_w and v_c as wind speed and velocity of the ocean current respectively. The central control agent collects them at the site of the accident.

$$\begin{aligned} s_w &= v_w \cdot t \\ s_c &= v_c \cdot t \end{aligned} \tag{5.2}$$

Fig. 5.4 Estimation of ship
drift direction and distance
accord to wind direction and
current direction

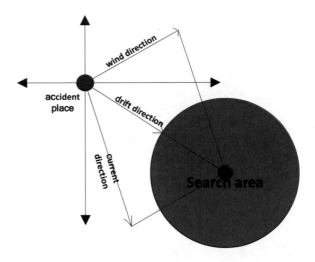

- After Determining the directions of the wind and current according to the latitude and longitude. As is shown in Fig. 5.4, we build a parallelogram whose diagonal line is the drift direction of the wrecked ship. The drift distance of the wrecked ship s_d can be calculated with the relative angle α and β. According to the actual situation, the reference point is regarded as the center, after which a certain length as a radius is selected to make a circle which is the range of search.

$$s_d = \sqrt{s_w^2 + s_c^2 - 2 \cdot s_w \cdot s_c \cdot cos\,(\pi - \alpha - \beta)} \qquad (5.3)$$

The multi-agent system adopts distributed computing, and each group of control agents completes path planning of search and rescue respectively. Firstly, the position between the group control agent and the target sea area is determined as the track area. And the weather information received by electronic charts and agents, barrier areas such as hurricanes and reefs will be set as restricted areas. Then select the midpoint of the adjacent bandgap as the path point. Finally, the algorithm can be used to calculate the shortest path from the beginning to the end.

5.2.2 Search Planning in the Wrecked Area

Every group's control agent calculates the search path ahead of time before entering the destruction zone from the alternate location. The search area is modeled as a rectangular area. And the ant path points are randomly generated in the region, the shortest traverse path is calculated by the improved ACO algorithm [9]. The shortest path is sent from the groups control agent to the search and rescue agents in the group. Each agent randomly enters the ant colony path point. The control agent is responsible for the communication connection between them, which is used to connect various information. Searchers then search clockwise according to the shortest path information received.

If a target is discovered during the search, the search agent will notify the control agent of the group [10]. The control agent issues a command that stops the process of search of all search agents, after which search agents are driven to the target area. In the meantime, the group control agent notifies other search groups to establish a temporary communication network in the area.

5.2.3 Establishment of Temporary Communication Network

Given the nature of sudden disasters such as shipwrecks, we often need more effective command and dispatch. In the absence of communication infrastructure, we rely on multi-agents such as USV as temporary communication nodes [11]. On the whole, the layered control structure is adopted to give full play to the edge computing ability of each agent. On the basis of mastering the information of each forwarding node's energy, channel capacity and so on, the PSO algorithm is applied to optimize the whole network and optimize the forwarding scheduling of the data packets in the node. Based on balancing network performance, network service requirements are further satisfied.

The temporary communication network is a multi-hop ad hoc network. In a reasonable range, by selecting a relay node with higher practical value, the hops number of data transmission can be reduced, and the time of data transmission can be accelerated. We make some assumptions on the key factors affecting the routing performance of the multi-hop network model [12]. Through the literature analysis, in the multi-hop transmission network model, the mobile planning of nodes in the network makes the routing strategy effective for the network. Meanwhile, although the multi-copy strategy improves parallelism and delivery speed of message routing, it also increases the consumption of network resources and network load to a certain extent, resulting in the degradation of routing performance [13]. Therefore, in the process of network design, we will use the edge computing ability of UAV and USV to analyze the mobile information of nodes in a specific geographical range and update the routing strategy in real-time by controlling the copy and hop count.

When a node in the network has enough node energy and certain buffer space, more relay nodes are used to forward multiple copies. Such metrics typically yield better QoS performance metrics (e.g., higher delivery rates and lower end-to-end time delay) [14]. However, in this temporary communication network model based on search and rescue sites, it is necessary to consider the stability of network nodes, such as node energy consumption and damage. Therefore, in this case, the network resources are strictly restricted, and the redundancy of message replication will occupy a large amount of bandwidth and spectrum resources. In addition, due to the energy is exhausted or there is no extra buffer space, the node will choose to

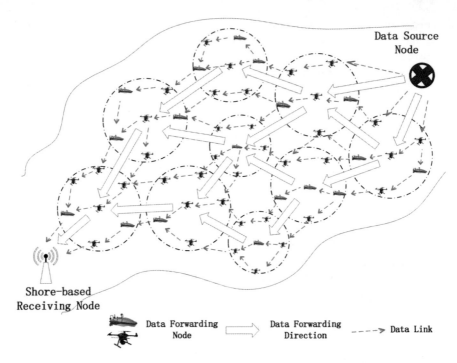

Fig. 5.5 Multi-hop network forwarding model based on multi-agent cooperation

stop the transmission and discard the message. Therefore, it is necessary to place restrictions on the number of relay nodes to avoid unnecessary redundancy.

In Fig. 5.5, data is offloaded to a base station on the bank by collaborative transmission between the original agent groups. The data is forwarded in a fixed direction, and the angle of the direction range is set to about 90. This measure successfully reduces the redundancy of messages in the network, saves network resource consumption, and allows messages to be forwarded to the onshore-based direction. Based on Fig. 5.5, we analyze how the next node's selection strategy is in the forwarding process. As shown in Fig. 5.6, we assume that N_X is the node that currently needs to make a route, and all the route nodes that can be selected for the next hop are defined as N_y. Within the communication range, there are other optional nodes such as N_A, N_B, N_T, etc. in the communication range pointing to the shore-based direction. The location of the relay node can be determined by the probability that the node appears at each location. From a statistical point of view, minimizing the variance of the desired direction is the best choice, so we need to get the desired direction of the node. All node position sets are represented by P, and the nodes are uniformly distributed within a certain range of positions, so the function of density probability is

$$f(x, y) = \frac{1}{S_{area}} = \frac{1}{\frac{1}{4}\pi R^2} \tag{5.4}$$

Fig. 5.6 Forwarding strategy to select next-hop relay node in multi-hop network

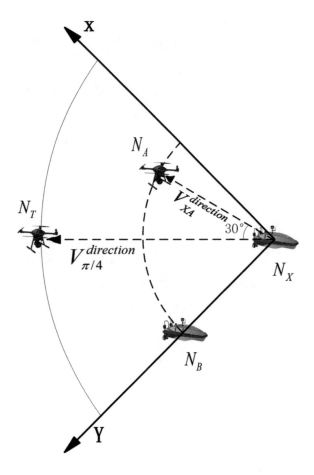

Among them, S_{area} represents the area of the region. The expected direction we can get from this is:

$$E[\theta] = \iint_S f_{X,Y}(x, y)\theta dS$$

$$= \int_0^{\frac{\pi}{2}} \int_0^R \frac{1}{\frac{1}{4}\pi R^2} r dr d\theta \qquad (5.5)$$

$$= \frac{1}{2}\theta^2 \Big|_0^{\frac{\pi}{2}} \cdot \frac{1}{\frac{1}{4}\pi R^2} \cdot \frac{1}{2}r^2 \Big|_0^R$$

$$= \frac{\pi}{4}$$

Each node in the network has to select the node closest to the expected direction to meet the actual task requirements. Therefore, according to the desired direction calculated by Eq. 5.5, the utility function selected by the next node can be defined. We use utility functions to reference the selection of the next node. The utility function we defined is:

$$
\begin{aligned}
U(N_X, N_A) = {} & \frac{1}{2R} D(N_X, N_A) \\
& + (1+\frac{\sqrt{2}}{2})(\cos\left\langle V_{XA}^{direction}, V_{\frac{\pi}{4}}^{direction}\right\rangle - \frac{\sqrt{2}}{2})
\end{aligned}
\tag{5.6}
$$

For all routing nodes N_y that can be selected for the next hop, we set

$$
\cos\left\langle V_{XX}^{direction}, V_{XY}^{direction}\right\rangle = \frac{\sqrt{2}}{2}
$$

As shown in Fig. 5.6, we use $V_{XA}^{direction}$ to indicate the direction from N_X to N_A. And we use $V_{\frac{\pi}{4}}^{direction}$ to represent the fan-shaped bisector. So in Eq. 5.6, $\frac{1}{2R} D(N_X, N_A)$ is the standard for evaluating distance. $(1+\frac{\sqrt{2}}{2})(\cos\left\langle V_{XA}^{direction}, V_{\frac{\pi}{4}}^{direction}\right\rangle - \frac{\sqrt{2}}{2})$ describes the similarity between $V_{XA}^{direction}$ and the expected optimal direction $V_{\frac{\pi}{4}}^{direction}$ discussed previously. So the scope of each part is $[0, \frac{1}{2}]$. And we get that $U \in [0, 1]$. So we introduce the two nodes N_T and N_X in Fig. 5.6. $V_{\frac{\pi}{4}}^{direction}$ is the direction of N_T, and the distance between it and N_X is R. So we can get the cosine of the angle formed by these two vectors

$$
\cos\left\langle V_{XT}^{direction}, V_{\frac{\pi}{4}}^{direction}\right\rangle = 1
$$

The most efficient value can be obtained from Eq. 5.6

$$
U(N_X, N_T) = \frac{1}{2R} R + (1 + \frac{\sqrt{2}}{2})(1 - \frac{\sqrt{2}}{2}) = 1
$$

Also, for node N_X

$$
D(N_X, N_X) = 0
$$

$$
\cos\left\langle V_{XT}^{direction}, V_{\frac{\pi}{4}}^{direction}\right\rangle = \cos\left\langle V_{XX}^{direction}, V_{XY}^{direction}\right\rangle = \frac{\sqrt{2}}{2}
$$

So we get the utility value

$$
U(N_X, N_X) = \frac{1}{2R} \cdot 0 + (1 + \frac{\sqrt{2}}{2})(\frac{\sqrt{2}}{2} - \frac{\sqrt{2}}{2}) = 0
$$

We can get $U(N_X, N_A)$ and $U(N_X, N_B)$ in the same way.

$$U(N_X, N_A) = \frac{1}{2R} \cdot \frac{R}{2} + (1 + \frac{\sqrt{2}}{2})(\frac{\sqrt{3}}{2} - \frac{\sqrt{2}}{2}) = 0.53$$

$$U(N_X, N_B) = \frac{1}{2R} \cdot \frac{R}{2} + (1 + \frac{\sqrt{2}}{2})(\frac{\sqrt{2}}{2} - \frac{\sqrt{2}}{2}) = 0.5$$

From this point of view, we can use Eq. 5.6 to calculate the utility value of node N_X to select the next-hop relay node within the communication range. For any relay node in the communication range, its relative utility value can be calculated. Since the utility value is related to the data transmission distance and antenna angle, we can estimate the specific time consumed by the data on the transmission link based on the utility value of these nodes. We can compare the utility value of nodes, and determine the data forwarding order according to the processing time and delivery time requirements of the data on the link as well as the utility value corresponding to the time, to obtain the maximum utilization efficiency of network resources.

5.3 The Improved Swarm Intelligence Algorithms

ACO and PSO methods have been applied to the fields of function optimization and engineering technology, and have achieved good results. In maritime search and rescue, it can be regarded as a TSP (Traveling Salesman Problem). Therefore, we mainly use ACO to plan the shortest path between start point and destination and finish the fastest traversal between the uniformly distributed coordinate points in the area. Meanwhile, PSO is used to forward and sort the packets in the node to maximize the utilization rate of network resources based on utility value. The ACO algorithm, which is combined with distributed computing, greedy search and feedback mechanism, can also fast search for the approximate optimal solution. Based on swarm intelligence theory, PSO is an optimization algorithm [15]. The algorithm realizes an efficient parallel search algorithm through the cooperation and competition between particles in the group [16]. The practice has proved that this method is suitable for dynamic and multi-objective optimization environment [17]. Therefore, it has good adaptability to the proposed multi-objective QoS optimization problem in the network.

5.3.1 Classical ACO Algorithm

In the process of applying ACO, ant colony path points are modeled by points on the path. Denote the number of initialized ants and the number of path points as m and n respectively. At this stage, set the first element of the taboo table $tabu$ of each ant to its current path point. In this context, the number of pheromones is equal on the paths, and $\tau_{ij}(0) = c$ is set. Next, each ant selects the next path points independently based on the number of remaining pheromones on the path and the heuristic information (distance between two path points). At t, the probability of the ant k transfer from the i point to the j point is $p_{ij}^k(t)$.

$$p_{ij}^k(t) = \begin{cases} \dfrac{[\tau_{ij}(t)]^\alpha \cdot [\eta_{ij}(t)]^\beta}{\sum\limits_{s \in J_k(i)} [\tau_{is}(t)]^\alpha \cdot [\eta_{is}]^\beta}, & s \in J_k(i) \\ 0, & other \end{cases} \tag{5.7}$$

where $J_k(i) = \{1, 2, \ldots, n\} - tabu_k$ defines that ant k is allowed to select set of path points in next step. Taboo table $tabu_k$ records the path points that ant k is passing currently. When all n path points are added to $tabu_k$, the ant k has completed a round trip. The path that ant k passes is a feasible solution to a TSP problem. η_{ij} in Eq. 5.7 is a heuristic factor representing the expectation of ants from point i to point j. η_{ij} usually takes the reciprocal of the distance between i and j. α represent the relative importance of the pheromone and β represents the expected heuristic factor. When all ants have completed one trip, the pheromone are updated according to the Eq. 5.8:

$$\tau_{ij}(t+n) = (1-\rho) \cdot \tau_{ij}(t) + \Delta\tau_{ij} \tag{5.8}$$

$$\Delta\tau_{ij} = \sum_{k=1}^m \Delta\tau_{ij}^k \tag{5.9}$$

Here, $\rho(0 < \rho < 1)$ is the pheromone evaporation coefficient on the path, $1 - \rho$ is the persistence coefficient of pheromone, $\Delta\tau_{ij}$ is the increment of pheromone on edge ij, $\Delta\tau_{ij}^k$ is the number of pheromone that the k ant left over on edge ij. If the ant k does not pass by the edge ij, then the value of $\Delta\tau_{ij}^k$ is 0. We define Q as the normal number, and L_k as the length of the path. $\Delta\tau_{ij}^k$ is expressed as:

$$\Delta\tau_{ij}^k = \begin{cases} \dfrac{Q}{L_k}, & when\ the\ ant\ k\ pass\ though\ the\ edge\ ij\ in\ this\ tour \\ 0, & other \end{cases} \tag{5.10}$$

5.3.2 The Improved ACO Algorithm

In the conventional ACO that was often used before, the parameters α and β stay still. As the algorithm iteratively progresses, the number of information on the path tends to average, which makes the ants lose sensitivity to the pheromone's change. This problem also causes local optimality and convergence oscillation during the iteration process. This section poses a method for adjusting α and β dynamically in Eq. 5.11:

$$\alpha = \begin{cases} \gamma\alpha, & \tau_{ij} \leq \tau \\ \alpha, & other \end{cases} , \beta = \begin{cases} \gamma\beta, & \tau_{ij} \geq \tau \\ \beta, & other \end{cases} \tag{5.11}$$

$$\gamma = \tau_{ij}(t)/\tau_{ij}(t+1) \tag{5.12}$$

where τ is the parameters in the experiment. Each time the ant completes the closed path, the two factors of pheromone, as well as expectation respectively, are adjusted by the set parameter γ. This promotion stabilizes the process of iterative convergence more, it also avoids the oscillation of local optimal. Figures 5.7 and 5.8 show experimental comparison results. We can see that in these two figures, the blue line is the shortest path length that the Dijkstra algorithm finds, with a total length of 218. The red line is the shortest path length found by ACO with a total length of 196.7815. After analyzing Figs. 5.7 and 5.8, the improved ACO iteration shows stability more, going to be stable at the fiftieth iteration. The improved ACO is utilized for the shortest path planning and the fastest traversal within the region.

5.3.3 Optimizing Packet Scheduling Order in Node

All particles in the PSO have a fitness value determined by the optimized function. The direction and distance of their flight are also determined by the speed of every particle. Then, the particles follow the current optimal particle search in the solution space. In a D-dimensional target search space, there is an assumption which is about N

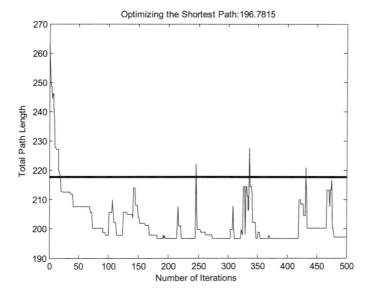

Fig. 5.7 Iterative process of standard ACO algorithm

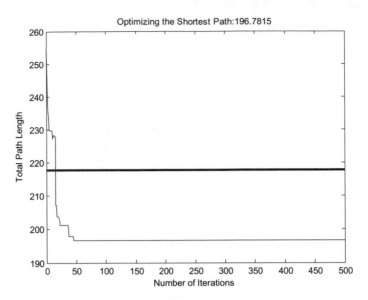

Fig. 5.8 Improved iterative process of ACO algorithm

particles forms a community, where the i-th particle is represented as a D-dimensional vector:

$$X_i = (x_{i1}, x_{i2}, \ldots, x_{iD}), \qquad i = 1, 2, \ldots, N \tag{5.13}$$

The flying speed of the i-th particle is also a D-dimensional vector, which is recorded as

$$V_i = (v_{i1}, v_{i2}, \ldots, v_{iD}), \qquad i = 1, 2, \ldots, N \tag{5.14}$$

The optimal position searched so far by i-th particle is called the individual extremum, recording as:

$$P_{best} = (p_{i1}, p_{i2}, \ldots, p_{iD}), \qquad i = 1, 2, \ldots, N \tag{5.15}$$

The optimal position searched so far by entire particle swarm is called the global extremum, recording as:

$$g_{best} = (g_1, g_2, \ldots, g_D) \tag{5.16}$$

When these two optimal values are found, the particles update their speed and position according to Eqs. 5.17 and 5.18 below:

$$v_{ij}(t+1) = v_{ij}(t) \\ + c_1 r_1(t)[p_{ij}(t) - x_{ij}(t)] + c_2 r_2(t)[p_{gj}(t) - x_{ij}(t)] \tag{5.17}$$

$$x_{ij}(t+1) = x_{ij}(t) + v_{ij}(t+1) \tag{5.18}$$

where c_1 and c_2 are learning factors; r_1 and r_2 are uniform random numbers in the range of $[0, 1]$, $i = 1, 2, \ldots, D$; v_{ij} is the velocity of the particle, $v_{ij} \in [-v_{max}, v_{max}]$, v_{max} is a constant and set by the user to limit the velocity of the particle. r_1 and r_2 are random numbers in the range of 0 and 1, which increase the randomness of particle flight.

To maintain a good convergence effect of the PSO, the ability of global search and local search reaches a balance point. Y.H. Shi introduces the inertia weight of w, the equation is as follows:

$$
\begin{aligned}
v_{ij}(t + 1) &= w \cdot v_{ij}(t) \\
&+ c_1 r_1(t)[p_{ij}(t) - x_{ij}(t)] + c_2 r_2(t)[p_{gj}(t) - x_{ij}(t)]
\end{aligned}
\tag{5.19}
$$

The inertia weight w indicates the degree of maintaining the original velocity and the larger w indicates the stronger the global convergence ability, and the weaker the local convergence ability, and vice versa. Besides, we can dynamically adjust w at any time during the search process and can make w larger at the beginning of the algorithm. As the search progresses, w can be gradually reduced. This allows the algorithm to detect more regions in the early stage, while in the later stage a fine search for extreme points. Its equation is as follows:

$$
w = w_{max} - \frac{(w_{max} - w_{min}) \cdot t}{T_{max}}
\tag{5.20}
$$

where T_{max} represents the maximum number of iterations, w_{min} represents the minimum inertia weight, w_{max} represents the maximum inertia weight, and t represents the current number of iterations.

5.4 Simulations of the Maritime Search and Rescue Mission Algorithms

The ant colony algorithm is used for shortest path planning and fastest traversing and has a successful implementation. At the same time, we use PSO to forward and sort packets in nodes. All the algorithms are simulated in MATLAB R 2014A environment. Set the simulation area according to different working conditions. During the simulation process, some obstacles to path planning were artificially set. The simulation software randomly generates the fastest coordinates in the region. For the utility value of each packet in the node, we also give the calculation method, which is reflected in the previous section. We set the primary forwarding node utility value and the auxiliary forwarding node utility value for each package, and use the particle swarm algorithm to balance the utility value with time.

5.4.1 Shortest Path Planning

After receiving the coordinate position of the target area, the group control agents will set an obstacle (dangerous sea area) between the start point and the endpoint in combination with information such as sea state, electronic chart and weather condition. To ensure safety, we choose the midpoint of the adjacent obstacle as the path point. Figures 5.9 and 5.10 demonstrate Shortest path in obstacle regions. The black area in the figure is an obstacle. S and T are the starting point and the middle point respectively. The red line in Fig. 5.9 is the shortest path planned by the ACO. The distance of the shortest path in Fig. 5.9 is 196.7815 compared with Fig. 5.8, while the result of the length of the general path calculated by the Dijkstra algorithm is 218. Figure 5.9 shows only the shortest path calculated by the ACO, and does not show the general path calculated by the Dijkstra algorithm. In Fig. 5.8, the shortest result of path length is the blue line found by the Dijkstra algorithm, while the red line is the shortest path length found by the ACO. Path calculated by ACO in Fig. 5.8 shows a dramatic shorter distance than the length calculated by the Dijkstra algorithm. Figure 5.10 shows the operation of the algorithm in the case of different obstacle areas with the same starting point. The simulation results show that the algorithm has a strong ability to handle different situations. In the simulation process, the number of ants is $m = 20$, the cycle number is set at 500, pheromone heuristic factor $\alpha = 3$, expected heuristic factor $\beta = 2$, pheromone evaporation coefficient $\rho = 0.8$.

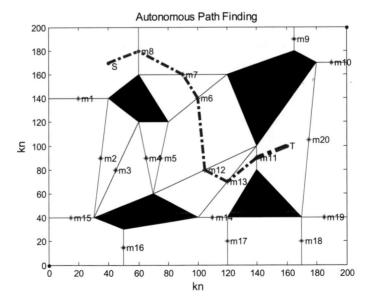

Fig. 5.9 Path simulation in the case of obstacle 1

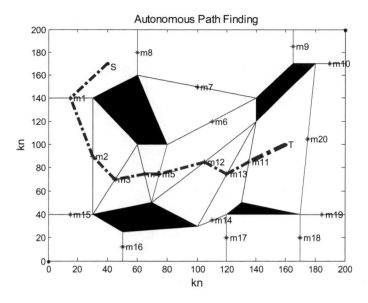

Fig. 5.10 Path simulation in the case of obstacle 2

5.4.2 The Fastest Traversal in the Region

At this time, activities of search and rescue which relays on large vessels mainly are combined with fan search, expanded square search, track search, parallel line saccade search, and other major search and rescue forms. Due to the search and rescue model proposed in this work, the agent can realize the autonomous planning of the search path in the region, as shown in Figs. 5.11 and 5.12. During the simulation, we set pheromone heuristic factor $\alpha = 1$, expected heuristic factor $\beta = 3$, the number of ants to $m = 50$, pheromone evaporation coefficient $\rho = 0.9$, the cycle number is set to 50, pheromone intensity coefficient $Q = 100$. In Fig. 5.11, we randomly select 80 points in the search area as the path points. From the traversal diagram, we can see that the traversal path is basically not repeated, and the path almost covers the entire area. In Fig. 5.12, we randomly traversed 200 points in this area. From the figure we can see that there are some paths overlapping, this area is covered by random points. During the actual search and rescue process, the number of random points is dynamically planned based on the performance of both requirements and equipment of the search and rescue on the agent. After entering the search area, the multi-agent group randomly enters a path point according to the search path of the group control agent plan and starts searching in a clockwise direction.

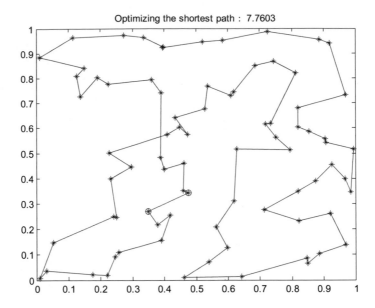

Fig. 5.11 The shortest traversal path in the case of 80 random points

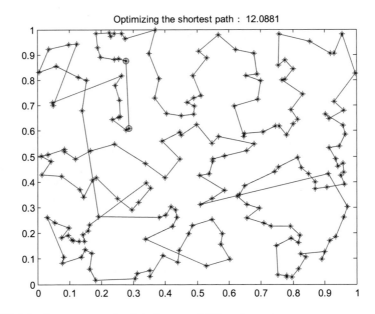

Fig. 5.12 The shortest traversal path in the case of 200 random points

5.4.3   Forwarding Sequence Scheduling of Data Packets Within a Node

When different types of packets arrive at the forwarding node, in a certain period, there must be multiple different types of packets in the node within a certain period, and packets of different priorities need to be forwarded. At this time, the proxy node uses PSO to calculate the forwarding order of the packets in the node according to the previously calculated packet forwarding time and the corresponding utility value. The maximum forwarding utility value to implement this forwarding. The forwarding time and the corresponding utility value of the packet are shown in Table 5.1. Based on these data, the forwarding order of the data packets when the maximum forwarding utility value is obtained is calculated, as shown in the sixth line of Table 5.1. At the same time, we use the PSO to calculate the iterative curve of the total utility value as shown in Fig. 5.13. Also, we pointed out the calculation of another set of data, as shown in Table 5.1 and Fig. 5.14.

As can be seen from Tables 5.1 and 5.2, we assume that the node has 9 packets to be forwarded at a certain time. They have different processing times, different forwarding time requirements, and corresponding utility values, etc. In the process of calculating the maximum utility value using the PSO, we select the number of group particles to be 100 and let them perform 20 iterations. Set the individual optimal learning factor C1 to 2 and the global optimal learning factor C2 to 2. At the same time, the inertia weight maximum value w_{max} is set to 0.9, and the inertia weight minimum value w_{min} is set to 0.4. At the same time, we deal with the boundary conditions and limit the particles to the feasible search space, avoiding the expansion and divergence of the population. The boundary velocity maximum value V_{max} is 4, and the minimum value V_{min} is -4.

Table 5.1 Packet forwarding time and utility value within the node 1

Packet number	1	2	3	4	5	6	7	8	9
Packet forwarding time on the link	0.3	0.4	0.1	0.2	0.6	0.1	0.4	0.7	0.5
Packet delivery time requirement	0.5	0.9	0.3	1.2	1.0	2.4	0.5	0.6	0.6
Deliver utility value on time	0.075	0.12	0.08	0.09	0.25	0.05	0.3	0.56	0.45
Overdue delivery utility value	0.05	0.09	0.04	0.075	0.18	0.03	0.15	0.4	0.2
Packet forwarding order	3	9	4	1	7	5	6	8	2

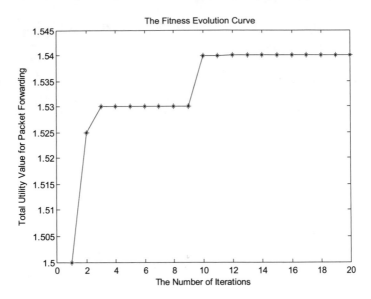

Fig. 5.13 Data forwarding total utility value evolution curve 1

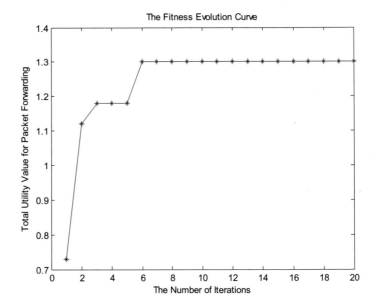

Fig. 5.14 Data forwarding total utility value evolution curve 2

Table 5.2 Packet forwarding time and utility value within the node 2

Packet number	1	2	3	4	5	6	7	8	9
Packet forwarding time on the link	0.4	0.3	0.6	0.5	0.2	0.3	0.1	0.5	0.7
Packet delivery time requirement	0.6	0.4	0.8	0.5	0.7	1.5	1.0	0.4	1.2
Deliver utility value on time	0.12	0.16	0.18	0.25	0.05	0.17	0.065	0.15	0.18
Overdue delivery utility value	0.10	0.12	0.14	0.02	0.11	0.045	0.10	0.13	0.06
Packet forwarding order	4	9	6	2	7	5	8	1	3

5.5 Summary

To realize the efficiency of maritime search and rescue, this chapter studies the maritime search and rescue network based on multi-agent cooperative communication, modeling and optimization. On the basis of edge computing and distributed control, the three respects about determination of location, planning of path as well as search of independent are assigned to each distributed group controller for distributed computing. Only a small amount of necessary information is communicated between the central controller and the group controller, which is suitable for QoS requirements at sea. Ant colony algorithm is used for heuristic greedy calculation. It helps the multi-agent in search and rescue activities to realize independent road search and cooperative search. At the same time, we explore the fastest way to cover the damaged area through the ant path point. In addition, the ad hoc network based on UAV and USV cluster proposed in this chapter has the characteristics of self-organization and adaptation. In the process of dynamic change of network topology, the data can be scheduled intelligently according to the forwarding strategy, and the network forwarding requirements can be completed. The joint search and rescue scheme and optimization strategy are established by using PSO algorithm. In the future work, we should explore the way to apply the advantages of different types of search and rescue to respect of personnel.

References

1. Pluhacek, M., Senkerik, R., Viktorin, A., Kadavy, T.: Exploring the shortest path in PSO communication network. In: Proceedings 2017 IEEE Symposium Series on Computational Intelligence (SSCI), Honolulu, HI, pp. 1–6 (2017)
2. Ruan, L., et al.: Energy-efficient multi-UAV coverage deployment in UAV networks: a game-theoretic framework. China Commun. **15**(10), 194–209 (2018)

3. Kim, G., Mahmud, I., Cho, I.: Self-recovery scheme using neighbor information for multi-drone ad hoc networks. In: Proceedings 2017 23rd Asia-Pacific Conference on Communications (APCC), Perth, WA, pp. 1–5 (2017)
4. Wang, T., Miao, Y., Ma, Q.: GMDSS automatic evaluation system. In: Proceedings 2010 International Conference on Logistics Engineering and Intelligent Transportation Systems, Wuhan, pp. 1–4 (2010)
5. Xu, H., Xu, W., Yang, Z., Shi, J., Chen, M.: Pilot reuse among D2D users in D2D underlaid massive MIMO systems. IEEE Trans. Veh. Technol. **67**(1), 467–482 (2018). Jan.
6. Nousiainen, J., Virtamo, J., Lassila, P.: Impact of multidirectional forwarding on the capacity of large wireless networks. IEEE Commun. Lett. **18**(2), 372–375 (2014). February
7. Fourmann, J. et al.: Wireless pressure measurement in air blast using PVDF sensors. In Proceedings: IEEE Sensors. Orlando, FL, pp. 1–3 (2016)
8. Xiao, Y., Hao, L., Ma, Z., Ding, Z., Zhang, Z., Fan, P.: Forwarding Strategy selection in Dual-Hop NOMA relaying systems. IEEE Commun. Lett. **22**(8), 1644–1647 (2018). Aug.
9. Zhang, J., Zhao, X., He, X.: A minimum resource neural network framework for solving multiconstraint shortest path problems. IEEE Trans. Neural Netw. Learn. Syst. **25**(8), 1566–1582 (2014). Aug.
10. Yanjie, C., Qiang, W.: A network specific information search system based on mobile agent. In Proceedings: Third Global Congress on Intelligent Systems. Wuhan, pp. 302–304 (2012)
11. Meng, X., Li, H., Cui, J.: Different strategies for differentially private histogram publication. J. Commun. Inf. Netw. **2**(3), 68C77 (2017)
12. Su, X., Hui, B., Chang, K.: Multi-hop clock synchronization based on robust reference node selection for ship ad-hoc network. J. Commun. Netw. **18**(1), 65–74 (2016). Feb.
13. Lin, C., Bi, Y., Zhao, H., Wang, Z., Wang, J.: Scheduling algorithms for time-constrained big-file transfers in the Internet of Vehicles. J. Commun. Inf. Netw. **2**(2), 126–135 (2017). Jun.
14. Zheng, C., Shan, Q., Zhang, H., Wang, Z.: On stabilization of stochastic cohen-grossberg neural networks with mode-dependent mixed time-delays and markovian switching. IEEE Trans. Neural Netw. Learn. Syst. **24**(5), 800–811 (2013). May
15. Wen, G., Zhang, Q., Wang, H., Tian, Q., Tao, Y.: An ant colony algorithm based on cross-layer design for routing and wavelength assignment in optical satellite networks. China Commun. **14**(8), 63–75 (2017). Aug.
16. Liu, X., Wei, Z.: Distributed computing system based on microprocessor cluster for wearable devices. In: Proceedings 2017 International Conference on Computer Network, Electronic and Automation (ICCNEA), Xi'an, pp. 66–71 (2017)
17. Du, W., Zhong, W., Tang, Y., Du, W., Jin, Y.: High-dimensional robust multi-objective optimization for order scheduling: a decision variable classification approach. IEEE Trans. Ind. Inf. **15**(1), 293–304 (2019). Jan.

Chapter 6
Conclusions and Future Directions

At present, although scholars have done a lot of research on maritime networks, there is still a certain distance to fully understand and utilize maritime communication networks. This monograph explores the problems existing in the maritime communication networks, including the establishment of a new network architecture based on SDN, the optimization of the offloading of computation-intensive tasks, and the application of maritime network multi-agent search and rescue, which provides some new ideas for solving mission-critical application driven intelligent maritime communications issues.

In this chapter, the main achievement of this monograph is summarized, further research directions are also pointed out. The brief contents are given as follows:

1. A software-defined maritime communication network framework is proposed. The framework integrates space, air, ground, and sea network deployment for joint transmission of data forwarding for intelligent and efficient transmission. In this framework, the scheduling strategy of the network resource is regarded as a joint optimization problem to improve throughput, balance delay and energy consumption, which combines channel state, cache state, and take multiple QoS as evaluation criteria. On this basis, an enhanced depth Q-learning method is proposed to solve the over-fitting and dimension curse problems when the service data volume grows on a large scale.
2. The issue about offloading tasks of computation for vessel terminals is investigated, which aims to minimize the energy consumption of vessel terminals and the execution delay of consumption tasks. Firstly, the strategy that whether the computation tasks should be offloaded to the cloud server is determined. Secondly, a server should be determined to derive the computation tasks. This work discusses the method to minimize the energy consumption of vessels while minimizing the execution delay of computation tasks with certain weights. To solve energy consumption and delay minimization problems, a multi-vessel computation offloading algorithm based on an improved Hungarian algorithm in the maritime MEC network.

T. Yang and X. Shen, *Mission-Critical Application Driven Intelligent Maritime Networks*, SpringerBriefs in Computer Science, https://doi.org/10.1007/978-981-15-4412-5_6

3. A mission-critical search and rescue network model based on multi-agent autonomous decision-making and task assignment is proposed under the proposed framework. The first level and the second level control agents use distributed edge computing to complete the tasks of path search and target location, respectively. In addition, after completing the target search, we use multiple agents to build a temporary communication network. At the same time, a communication forwarding strategy based on the multi-agent joint network is designed, which takes full advantage of the load capacity of the network. The network model strategy can integrate the search and rescue resources, independently analyze the scene situation, and finally provide a complete search and rescue process.

The existing work, as the cornerstone of the future work, is conducive to the next step. The future research direction will refine the functions faced by the maritime communication network. While making full use of the existing network, we will study the future maritime communication network enabled by AI. For example, how to apply AI algorithms in the maritime communication networks, and what roles do these algorithms play in the system.